My Book

This book belongs to

Name: _____

Copy right © 2019 MATH-KNOTS LLC

All rights reserved, no part of this publication may be reproduced, stored in any system or transmitted in any form, or by any means, electronic, mechanical, photocopying, recording, or otherwise without the written permission of MATH-KNOTS LLC.

Cover Design by :
Gowri Vemuri

First Edition :
April , 2021

Author :
Gowri Vemuri

Edited by :
Raksha Pothapragada
Ritvik Pothapragada

Questions: mathknots.help@gmail.com

NOTE : These tests are neither affiliated nor sponsored or endorsed by any organization.

Dedication

This book is dedicated to:
My Mom, who is my best critic, guide and supporter.
To what I am today, and what I am going to become tomorrow,
is all because of your blessings, unconditional affection and support.

This book is dedicated to the
strongest women of my life,
my dearest mom
and
to all those moms in this universe.

G.V.

Visit www.a4ace.com

Also available more time based practice tests on subscription
Subscribe to Math-Knots you tube channel for concept videos
E-mail us: mathknots.help@gmail for more concept based videos with a proof of purchase.

Topic and page numbers

Index

Preface	1 - 12
Notes	13 - 82
Fractions #1	83 - 88
Fractions #2	89 - 92
Fractions #3	93 - 96
Decimals #4	97 - 99
Decimal #5	100 - 102
Decimal #6	103 - 105
Integer Multiplication #7	106 - 108
Scientific Notation #8	109 - 113
GCF #9	114 - 116
GCF Monomials #10	117 - 120
LCM Numbers #11	121 - 123
LCM Monomials #12	124 - 126

Index

Topic	Pages
Order of Operations #13	127 - 129
Verbal Expressions #14	130 - 132
Verbal Expression Equation #15	133 - 137
Monomials #16	138 - 140
Inequalities #17	141 - 144
Evaluate Expressions #18	145 - 147
Evaluate Expressions #19	148 - 150
Solve for X #20	151 - 154
Absolute Value #21	155 - 157
Absolute Value #22	158 - 162
Proportions #23	163 - 166
Proportions #24	167 - 170
Percent Discount #25	171 - 174

Topic and page numbers

Index

Topic	Pages
Percent MarkUp #26	175 - 178
Percent Tax #27	179 - 182
Percent Change #28	183 - 186
Radicals #29	187 - 190
Radicals #30	191 - 194
Radicals #31	195 - 198
Radicals #32	199 - 203
Radicals #33	204 - 207
Radicals #34	208 - 211
Triangles #35	212 - 214
Rectangle-Square #36	215 - 217
Parallelogram Trapezium #37	218 - 220
Circle - Area - Circumference #38	221 - 224

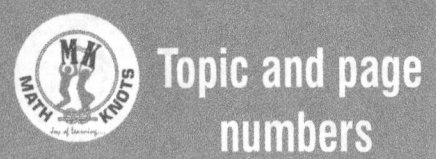

Index

Topic and page numbers

Topic	Pages
Volume-Sphere #39	225- 228
Volume-Rectangle Square prisms #40	229 - 233
Volume - Cone - Cylinder #41	234 - 238
Additive Inverse #42	239 - 242
Multiplicative Inverse #43	243- 246
Rational - Irrational Numbers #44	247 - 249
Exponents #45	250 - 253
Slope - 2 points #46	254 - 258
Slope intercept form #47	259 - 263
Slope Graph #48	264 - 267
Straight line slope #49	268- 270
Parallel line slope #50	271- 273
Perpendicular line slope #51	274 - 276

Index

Topic	Pages
Missing angle #52	277 - 279
Missing angle #53	280 - 283
Reflection #54	284 - 287
Rotation #55	288 - 291
Translation #56	292 - 295
One step word problems #57	296 - 298
System of equations #58	299 - 302
System of equations #59	303 - 306
Line graph #60	307 - 316
Scatter plot #61	317 - 326
Stem leaf plot #62	327 - 332
Dot plot #63	333 - 337
Pie chart #64	338 - 348
Answer keys	349 - 370

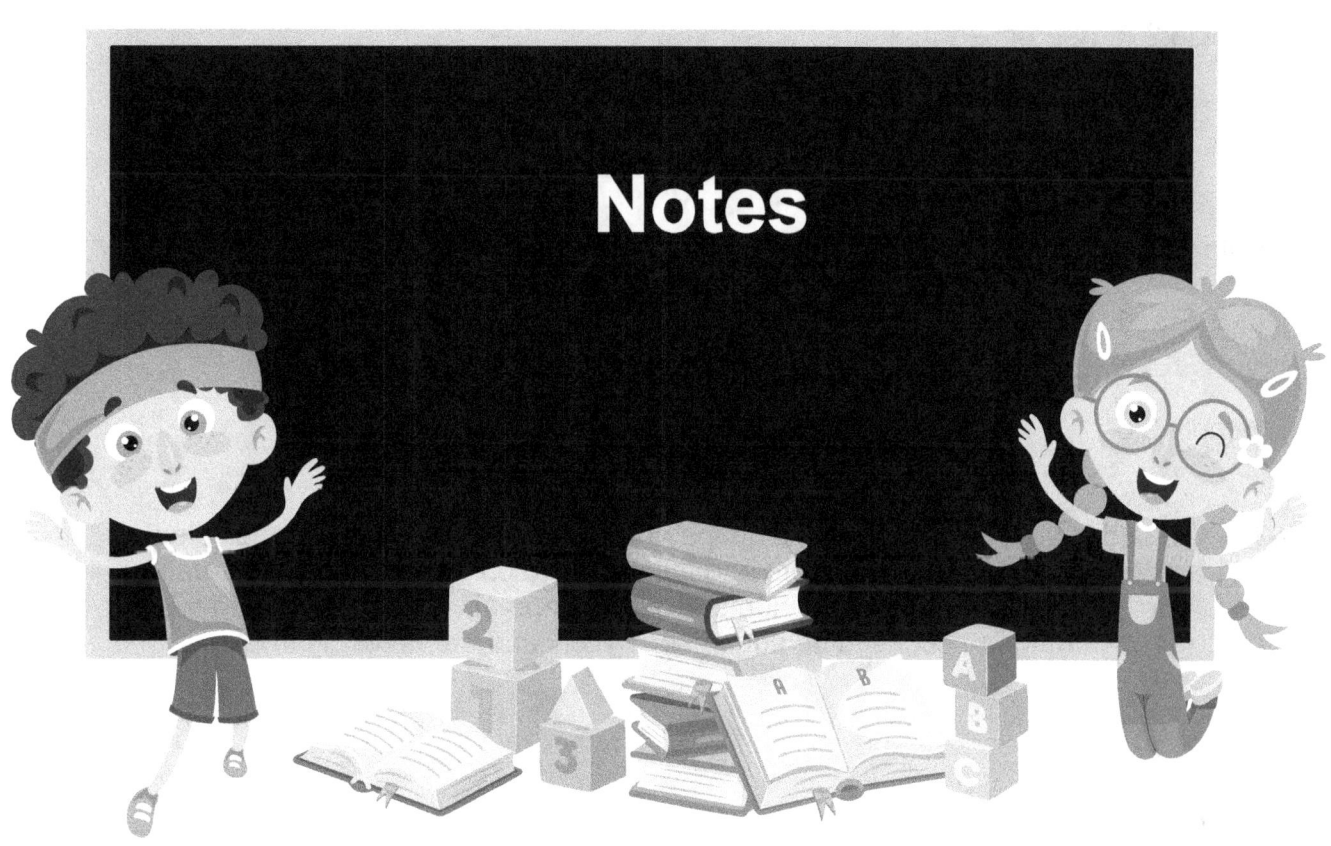

Fractions Notes

Fractions are part of a whole. It is also an expression representing quotient of two quantities.

Example 1 : $\frac{2}{4}, \frac{1}{4}$

Fraction can also be represented as ratios.

Example 2 : 1 : 2 or $\frac{1}{2}$

3 : 7 of $\frac{3}{7}$

Adding simple fractions, follow the below steps :

#1 : Fractions with like denominators can be added by adding their numerators.

Example 3 : $\frac{3}{5} + \frac{1}{5} = \frac{3+1}{5} = \frac{4}{5}$

#3 : To add fractions with unlike denominators, convert the fractions to equivalent fractions with like denominators and follow #1. To convert them into equivalent fractions you can multiply the numerator and denominator with a common factor for each of the fractions to be added separately and then add the fractions.

Method 1 :

Example 4 : $\frac{2}{5} + \frac{1}{2}$

Step 1 : $\frac{2}{5} = \frac{2 \times 2}{5 \times 2} = \frac{4}{10}$

Step 2 : $\frac{1}{2} = \frac{1 \times 5}{2 \times 5} = \frac{5}{10}$

Step 3 : $\frac{2}{5} + \frac{1}{2} = \frac{4}{10} + \frac{5}{10} = \frac{4+5}{10} = \frac{9}{10}$

$\frac{2}{5} + \frac{1}{2} = \frac{9}{10}$

FRACTIONS

Notes

Another method to add fractions with unlike denominators is by finding the **Least Common Multiple** (LCM) of the denominators and then follow the steps as described below in same order.

Method 2 :

Example 5 : $\frac{5}{12} + \frac{1}{8}$

Step 1 :   Find the LCM of the denominators 12 , 8

$$\begin{array}{c|c} 2 & 12 , 8 \\ 2 & 6 , 4 \\ 3 & 3 , 2 \\ 2 & 1 , 2 \\ & 1 , 1 \end{array}$$ LCM = 2 X 2 X 3 X 2 = 24

Step 2 :   Converting $\frac{5}{12}$ into an equivalent fraction with the denominator as 24 (LCM).

Divide the LCM value with the number in the denominator of the fraction to obtain the common factor.

$$12 \overline{)\begin{array}{r} 2 \\ 24 \\ -24 \\ \hline 0 \end{array}}$$

Common factor obtained is 2

$$\frac{5}{12} = \frac{5 \times 2}{12 \times 2} = \frac{10}{24} \; ; \; \frac{5}{12} = \frac{10}{24}$$

Step 3 :   Converting $\frac{1}{8}$ into an equivalent fraction with the denominator as 24 (LCM).

Divide the LCM value with the number in the denominator of the fraction to obtain the common factor.

$$8 \overline{)\begin{array}{r} 3 \\ 24 \\ -24 \\ \hline 0 \end{array}}$$

Common factor obtained is 3

$$\frac{1}{8} = \frac{1 \times 3}{8 \times 3} = \frac{3}{24} \; ; \; \frac{1}{8} = \frac{3}{24}$$

FRACTIONS

Notes

Step 4 : Substitute the equivalent fractions obtained in step 2 and 3.

$$\frac{5}{12} + \frac{1}{8} = \frac{10}{24} + \frac{3}{24}$$

Step 5 : The denominators of both fractions are same (Like denominators). Add the numerators.

$$\frac{10+3}{24} = \frac{13}{24}$$

Note : To add more than two fractions, repeat step 2 or step 3 for each of the fractions to convert them into equivalent fractions and then proceed with step 4 and 5.

$$\frac{5}{12} + \frac{1}{8} = \frac{13}{24}$$

Subtracting simple fractions, follow the below steps :

#1 : Fractions with like denominators can be subtracted by subtracting their numerators.

Example 6 : $\frac{3}{5} - \frac{1}{5} = \frac{3-1}{5} = \frac{2}{5}$

#2 : To subtract fractions with unlike denominators, convert the fractions to equivalent fractions with like denominators and follow #1. To convert them into equivalent fractions you can multiply the numerator and denominator with a common factor for each of the fractions to be subtracted separately and then subtract the fractions.

Method 1 :

Example 7 : $\frac{3}{5} - \frac{1}{2}$

Step 1 : $\frac{3}{5} = \frac{3 \times 2}{5 \times 2} = \frac{6}{10}$

Step 2 : $\frac{1}{2} = \frac{1 \times 5}{2 \times 5} = \frac{5}{10}$

Step 3 : $\frac{3}{5} - \frac{1}{2} = \frac{6}{10} - \frac{5}{10} = \frac{6-5}{10} \quad \frac{1}{10}$

$$\frac{3}{5} - \frac{1}{2} = \frac{1}{10}$$

FRACTIONS

Notes

Another method to subtract fractions with unlike denominators is by finding the **Least Common Multiple** (LCM) of the denominators and then follow the steps as described below in same order.

Method 2 :

Example 8 : $\frac{5}{12} - \frac{1}{8}$

Step 1 :   Find the LCM of the denominators 12 , 8

```
2 | 12 , 8
2 |  6 , 4      LCM = 2 X 2 X 3 X 2 = 24
3 |  3 , 2
2 |  1 , 2
     1 , 1
```

Step 2 :   Converting $\frac{5}{12}$ into an equivalent fraction with the denominator as 24 (LCM).

Divide the LCM value with the number in the denominator of the fraction to obtain the common factor.

$$12\overline{)24} \quad \text{quotient } 2, \text{ remainder } 0$$

Common factor obtained is 2

$$\frac{5}{12} = \frac{5 \times 2}{12 \times 2} = \frac{10}{24} \quad ; \quad \frac{5}{12} = \frac{10}{24}$$

Step 3 :   Converting $\frac{1}{8}$ into an equivalent fraction with the denominator as 24 (LCM).

Divide the LCM value with the number in the denominator of the fraction to obtain the common factor.

$$8\overline{)24} \quad \text{quotient } 3, \text{ remainder } 0$$

Common factor obtained is 3

$$\frac{1}{8} \quad \frac{1 \times 3}{8 \times 3} \quad \frac{3}{24} \quad \frac{1}{8} \quad \frac{3}{24}$$

FRACTIONS

Notes

Step 4 : Substitute the equivalent fractions obtained in step 2 and 3.

$$\frac{5}{12} - \frac{1}{8} = \frac{10}{24} - \frac{3}{24}$$

Step 5 : The denominators of both fractions are same (Like denominators). Subtract the numerators.

$$\frac{10 - 3}{24} = \frac{7}{24}$$

Note : To Subtract more than two fractions, repeat step 2 or step 3 for each of the fractions to convert them into equivalent fractions and then proceed with step 4 and 5.

$$\frac{5}{12} - \frac{1}{8} = \frac{7}{24}$$

Adding Mixed numbers :

#1 : Add the whole numbers together.

#2 : Add the fractional parts together. (Find a common denominator if necessary) (Follow the steps as described in the previous pages)

#3 : Write the whole number obtained from step 1 and the fraction obtained from step 2.

#4 : If the fractional part is an improper fraction, change it to a mixed number. Add the whole part of the mixed number to the original whole numbers. Rewrite the fraction in the lowest possible value.

Example 9 : $3\frac{2}{7} + 5\frac{1}{7}$

Step 1 : Add the whole part 3 from $3\frac{2}{7}$ to the whole part 5 from $5\frac{1}{7}$

$3 + 5 = 8$

Step 2 : Add the fractional part $\frac{2}{7}$ from $3\frac{2}{7}$ to the fractional part $\frac{1}{7}$ from $5\frac{1}{7}$

$$\frac{2}{7} + \frac{1}{7} = \frac{2 + 1}{7} = \frac{3}{7}$$

$$3\frac{2}{7} + 5\frac{1}{7} = 8\frac{3}{7}$$

 **FRACTIONS**

Example 10 : $4\frac{1}{6} + 5\frac{5}{6}$

Step 1 :   Add the whole part 4 from $4\frac{1}{6}$ to the whole part 5 from $5\frac{5}{6}$
          $4 + 5 = 9$

Step 2 :   Add the fractional part $\frac{1}{6}$ from $4\frac{1}{6}$ to the fractional part $\frac{5}{6}$ from $5\frac{5}{6}$
          $\frac{1}{6} + \frac{5}{6} = \frac{1+5}{6} = \frac{6}{6}$ (Numerators are added for fractions from like denominators)

Step 3 : $4\frac{1}{6} + 5\frac{5}{6} = 9\frac{6}{6} = 10$

          ($\frac{6}{6}$ = one whole part, add this one whole part to 9 making it equal to 10)

Example 11 : $7\frac{1}{2} + 5\frac{3}{20}$

Step 1 :   Add the whole part 7 from $7\frac{1}{2}$ to the whole part 5 from $5\frac{3}{20}$
          $7 + 5 = 12$

Step 2 :   Add the fractional part $\frac{1}{2}$ from $7\frac{1}{2}$ to the fractional part $\frac{3}{20}$ from $5\frac{3}{20}$

          $\frac{1}{2} + \frac{3}{20}$ (The fractions has unlike denominators)

          Finding the LCM of 2 and 20

          ```
          2 | 2 , 20
          2 | 1 , 10     LCM of 2 and 20 = 2 X 2 X 5 = 20
          5 | 1 , 5
              1 , 1
          ```

Step 3 :   Let's make $\frac{1}{2}$ as an equivalent fraction with a denominator of 20.

          ```
              10
          2) 20      Common factor is 10
            -20
             ___
              0
          ```

FRACTIONS

Notes

$$\frac{1}{2} = \frac{1 \times 10}{2 \times 10} = \frac{10}{20}$$

Step 4 : The fractional part $\frac{3}{20}$ of $5\frac{3}{20}$ has the same common denominator as 20.

We do not need to convert $\frac{3}{20}$ into another equivalent fraction.

Remember : The fractions can vary from problem to problem and students need to follow step 3 for all the fractional parts to convert them to equivalent fractions.

Step 5 : $\frac{1}{2} + \frac{3}{20} = \frac{10}{20} + \frac{3}{20} = \frac{10+3}{20} = \frac{13}{20}$

Step 6 : $7\frac{1}{2} + 5\frac{3}{20} = 8\frac{13}{20}$

Subtracting Mixed numbers :

#1 : Subtract the whole numbers together.

#2 : Subtract the fractional parts together. (Find a common denominator if necessary) (Follow the steps as described in the previous pages)

#3 : Write the whole number obtained from step 1 and the fraction obtained from step 2.

#4 : If the fractional part is an improper fraction, change it to a mixed number.
Add the whole part of the mixed number to the whole number obtained in step 1.
Rewrite the fraction in the lowest possible value.

Example 12 : $9\frac{2}{7} - 5\frac{1}{7}$

Step 1 : Subtract the whole part 5 from $5\frac{1}{7}$ from the whole part 9 from $3\frac{2}{7}$

$9 - 5 = 4$

Step 2 : Subtract the fractional part $\frac{1}{7}$ from $5\frac{1}{7}$ from the fractional part $\frac{2}{7}$ from $3\frac{2}{7}$

$\frac{2}{7} - \frac{1}{7} = \frac{2-1}{7} = \frac{1}{7}$

Step 3 : $9\frac{2}{7} - 5\frac{1}{7} = 4\frac{1}{7}$

FRACTIONS

Notes

Example 13 : $7\frac{5}{6} - 5\frac{1}{6}$

Step 1 : Subtract the whole part 5 from $5\frac{1}{6}$ from the whole part 7 from $7\frac{1}{6}$

7 - 5 = 2

Step 2 : Subtract the fractional part $\frac{1}{6}$ from $5\frac{1}{6}$ to the fractional part $\frac{5}{6}$ from $7\frac{5}{6}$

$\frac{5}{6} - \frac{1}{6} = \frac{5-1}{6} = \frac{4}{6}$ (Numerators are added for fractions from like denominators)

Step 3 : $7\frac{5}{6} - 5\frac{1}{6} = 2\frac{4}{6} = 2\frac{2}{3}$

($\frac{4}{6} = \frac{2}{3}$ Equivalent fractions)

Example 14 : $7\frac{1}{2} - 5\frac{3}{20}$

Step 1 : Subtract the whole part 5 from $5\frac{3}{20}$ from the whole part 7 from $7\frac{1}{2}$

7 - 5 = 2

Step 2 : Subtract the fractional part $\frac{3}{20}$ from $5\frac{3}{20}$ to the fractional part $\frac{1}{2}$ from $7\frac{1}{2}$

$\frac{3}{20} - \frac{1}{2}$ (The fractions has unlike denominators)

Finding the LCM of 2 and 20

```
2 | 2 , 20
2 | 1 , 10     LCM of 2 and 20 = 2 X 2 X 5 = 20
5 | 1 , 5
    1 , 1
```

Step 3 : Let's make $\frac{1}{2}$ as an equivalent fraction with a denominator of 20.

```
    10
2) 20       Common factor is 10
  -20
   ---
    0
```

FRACTIONS

Notes

$$\frac{1}{2} = \frac{1 \times 10}{2 \times 10} = \frac{10}{20}$$

Step 4: The fractional part $\frac{3}{20}$ of $5\frac{3}{20}$ has the same common denominator as 20.

We do not need to convert $\frac{3}{20}$ into another equivalent fraction.

Remember: The fractions can vary from problem to problem and students need to follow step 3 for all the fractional parts to convert them to equivalent fractions.

Step 5: $\frac{1}{2} - \frac{3}{20} = \frac{10}{20} - \frac{3}{20} = \frac{10-3}{20} = \frac{7}{20}$

Step 6: $7\frac{1}{2} - 5\frac{3}{20} = 2\frac{7}{20}$

Example 15: $5\frac{1}{7} - 3\frac{3}{7}$

Step 1: Subtract the whole part 3 from $3\frac{3}{7}$ from the whole part 5 from $5\frac{1}{7}$

$5 - 3 = 2$

Step 2: Subtract the fractional part $\frac{3}{7}$ from $3\frac{3}{7}$ from the fractional part $\frac{1}{7}$ from $5\frac{1}{7}$

We cannot subtract $\frac{3}{7}$ from $\frac{1}{7}$

So we need to rewrite the fraction $5\frac{1}{7}$

$5\frac{1}{7} = 4\frac{8}{7}$ (Remember in this fraction one whole part equals to seven. so when we take one whole part into the fraction form we need to add 7 to the value in the numerator which equals to 7 + 1 = 8)

Step 3: Repeat step 1

$5\frac{1}{7} - 3\frac{3}{7} = 4\frac{8}{7} - 3\frac{3}{7}$

Subtract the whole part 3 from $3\frac{3}{7}$ from the whole part 4 from $4\frac{8}{7}$

$4 - 3 = 1$

FRACTIONS

Notes

Step 4 : Subtract the fractional part $\frac{3}{7}$ from $3\frac{3}{7}$ from the fractional part $\frac{8}{7}$ from $4\frac{8}{7}$

$$\frac{8}{7} - \frac{3}{7} = \frac{8-3}{7} = \frac{5}{7}$$

Step 5 : $5\frac{1}{7} - 3\frac{3}{7} = \frac{5}{7}$

Multiplying Fractions :

#1 : Verify if the fractions are in lowest possible values. If not convert them into lowest possible values.

#2 : Using cross simplification method simplify the fractions, meaning a numerator can be simplifies with a denominator only and vice versa.

#3 : Do not cross simplify numerator with a numerator value and denominator with a denominator value

#4 : Multiply the numerator with the remaining numerator values and the denominator with the denominator values

Remember : "Top times the top over the bottom times the bottom".
All the answers must be written in simplest form.

Example 16 : $\frac{6}{15} \times \frac{3}{10}$

$$\frac{\overset{2}{\cancel{6}}}{\underset{5}{\cancel{15}}} \times \frac{3}{10} = \frac{\overset{1}{\cancel{2}}}{5} \times \frac{3}{\underset{5}{\cancel{10}}} = \frac{1}{5} \times \frac{3}{5}$$

$$= \frac{1 \times 3}{5 \times 5} = \frac{3}{25}$$

FRACTIONS

Multiplying Mixed Numbers :

#1 : To multiply mixed numbers, convert them to improper fractions.

Converting mixed number to improper fractions :
Multiply the denominator of the fraction to the whole part and then add the product to the numerator.

Example 17 : $2\frac{1}{3} = \frac{2 \times 3 + 1}{3} = \frac{6+1}{3} = \frac{7}{3}$

#2 : Verify if the fractions are in lowest possible values. If not convert them into lowest possible values.

#3 : Using cross simplification method simplify the fractions, meaning a numerator can be simplifies with a denominator only and vice versa.

#4 : Do not cross simplify numerator with a numerator value and denominator with a denominator value

#5 : Multiply the numerator with the remaining numerator values and the denominator with the denominator values

Remember : "Top times the top over the bottom times the bottom".
All the answers must be written in simplest form.
All improper fractions must be change back to mixed numbers.

Note : MULTIPLICATION CAN BE WRITTEN WITH THE SYMBOLS X OR . IN BETWEEN, .

FRACTIONS

Dividing Fractions :

#1 : To divide fractions, convert the division problem into a multiplication problem.
Do this by multiplying the first fraction by the reciprocal of the second fraction.
In other words convert the division to multiplication and interchange the numerator and denominator of the second fraction.

Remember : "When two fractions we divide, flip the second and multiply."

Note : Don't forget to check for "cross simplification" when multiplying.
All answers must be written in simplest form.
All improper fractions must be change back to mixed numbers.

Example 18 : $\dfrac{6}{15} \div \dfrac{3}{10}$

$$\dfrac{6}{15} \div \dfrac{3}{10} = \dfrac{\overset{2}{\cancel{6}}}{\underset{5}{\cancel{15}}} \times \dfrac{10}{3} = \dfrac{2}{\underset{1}{\cancel{5}}} \times \dfrac{\overset{2}{\cancel{10}}}{3} = \dfrac{2}{1} \times \dfrac{2}{3}$$

$$= \dfrac{2 \times 2}{1 \times 3} = \dfrac{4}{3}$$

Dividing Mixed Numbers :

#1 : To divide mixed fractions, first change them to improper fractions.

#2 : To divide fractions, convert the division problem into a multiplication problem.
Do this by multiplying the first fraction by the reciprocal of the second fraction.
In other words convert the division to multiplication and interchange the numerator and denominator of the second fraction.

Remember : "When two fractions we divide, flip the second and multiply."

Note : Don't forget to check for "cross simplification" when multiplying.
All answers must be written in simplest form.
All improper fractions must be change back to mixed numbers.

Example 19 : $2\dfrac{6}{15} \div 5\dfrac{3}{10}$

$$2\dfrac{6}{15} \div 5\dfrac{3}{10} = \dfrac{2 \times 15 + 6}{15} \div \dfrac{5 \times 10 + 3}{10} = \dfrac{30 + 6}{15} \div \dfrac{50 + 3}{10}$$

FRACTIONS

$$= \frac{30+6}{15} \div \frac{50+3}{10}$$

$$= \frac{36}{15} \div \frac{53}{10}$$

$$= \frac{36}{15} \times \frac{10}{53} \quad \text{(Remember when we change division to multiplication, flip the second fraction)}$$

$$= \frac{\overset{12}{\cancel{36}}}{\underset{5}{\cancel{15}}} \times \frac{10}{53}$$

$$= \frac{12}{\underset{1}{\cancel{5}}} \times \frac{\overset{2}{\cancel{10}}}{53}$$

$$= \frac{12}{1} \times \frac{2}{53}$$

$$= \frac{12 \times 2}{1 \times 53}$$

$$= \frac{24}{53}$$

$$2\frac{6}{15} \div 5\frac{3}{10} = \frac{24}{53}$$

DECIMALS

Decimals notes

The word standard means regular. Numbers in standard form are whole numbers or natural numbers.

Example : The number "six hundred twenty five" in standard form is 625.

To name a decimal from its standard form, follow these steps :

1. Name the number in front of the decimal. (Do not include the word "and").

2. The word "and" is used for the decimal point.

3. The number in the decimal part is similar to the number in front of the decimal.

4. Name the last place value given (of the digit farther to the right).

Rounding Decimals to a Given Place Value:

To round a decimal number to a given place value, look at the digit to the right of the desired place value and follow the rounding rules:

"5 and above, give it a shove! 4 or below, leave it alone!"

(The number in the desired place value gets "bumped up" to the next consecutive value if the digit to the right of it is 5 or more. The number in the desired place value does not change if the digit to the right of it has a value of 5 or less.)

The purpose of rounding is to provide *an* estimate and get an approximate value. Rounding involves losing some accuracy.

Example : If 5,953 people attend a Soccer game. We can approximately say 6,000 people watched the game.

Comparing and Ordering Decimals:

To order the given decimals, compare the digits of all decimals according to their place value.

Tip : Line the numbers up vertically according to their place value and compare from the left to the right. When comparing like place values from the left, the number with the higher digit is the larger number.

Also, added zeros to the right of a decimal number does not change the value of the decimal.

For example, 0.8 = 0.80 = 0.800 = 0.8000 = 0.80000

DECIMALS

Naming/Reading Decimal :

Read the number to the left of the decimal. Say the word "and" for the decimal place. Read the number to the right of the decimal as you would read it if the number were on the left. End the name with the place value of the digit that is furthest to the right.

Converting Between Fractions and Decimals :

Fractions and decimals represent a certain "part of a whole", all fractions can be written as decimals and terminating/repeating decimals can be written as fractions.

Multiplying Decimals :

To multiply decimals, first ignore the decimals and simply multiply the digits. Then count the total number of spaces from the right to the decimal (or digits to the right of the decimal) of both numbers; place the decimal that number of spaces from the right into your answer.

Dividing Decimals:

Divide the decimals as a whole numbers ignoring the decimals.

1. Count the number of digits of the dividend after the decimal.
2. Count the number of digits of the divisor.
3. Subtract the number obtained in step 1 from the number obtained instep 2.
4. Count the number of digits from the right to the number obtained in step 3.
5. Place the decimal point.

INTEGERS

Integers notes

Integers are a group, or set of numbers that consist of "whole numbers and their opposites"

1. Natural numbers and whole numbers are subset of integers.

2. The set does not include fractions or decimals.

3. The set includes positive and negative numbers.

4. Integers include : $-\infty$....., -5, -4, -3, -2, -1, 0, 1, 2, 3, 4, 5 ...... $+\infty$

5. Integers greater than zero are called positive integers.

6. Integers less than zero are called negative integers.

7. Zero is neither negative nor positive.

8. Negative integers are the numbers to the left of 0.
   Example : -5, -4, -3

9. Negative numbers have a negative (-) sigh in front of the number.

10. Positive integers are the numbers to the right of 0.

11. Positive numbers do not require the + sign in front.

12. If a number has no sign, it is a positive number.
    Example : 2, 3, 4, 10, 20

13. Negative numbers are frequently used in measurements.
    Example :   To measure temperatures, depth etc
    4^0 C below zero degree celsius is represented as -4^0
    100 ft below sea level is represented as -100 ft.

14. Arrows on a number line represent the numbers continuing for ever.

15. Positive numbers are represented on the right side of zero on the number line.

16. Negative numbers are represented on the left side of zero on the number line.

17. Number are placed at equal intervals on the number line. Not necessarily one unit.

INTEGERS

Integers can be represented on a number line as below

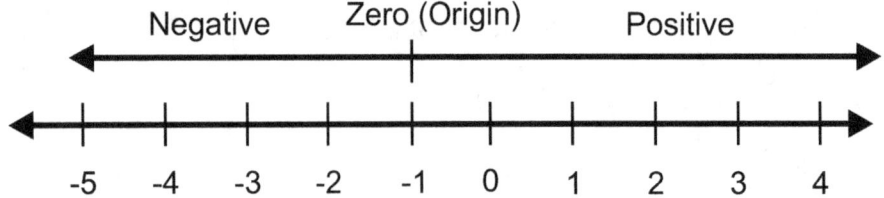

Absolute value :

The number line can be used to find the absolute value. The absolute value of an integer is the distance the number is from zero on the number line.

The absolute value of 2 is 2. Using the number line, 2 is a distance of 2 to the right of zero. The absolute value of -2 is also 2. Again using the number line, the distance from -2 to zero is 2. A measure of distance is always positive.

The symbol for absolute value of any number, x, is | x |.

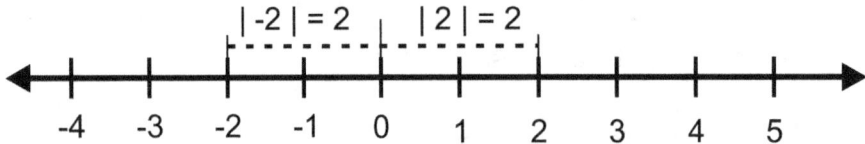

Opposite integers :

The opposite of an integer is the number that is at the same distance from zero in the opposite direction. Every integer has an opposite value, but the opposite of zero is itself.

The opposite of -4 is 4 because it is located the same distance from zero as 4 is, but in opposite direction.

INTEGERS

Notes

Adding integers using a number line :

The number line is visual representation to understand the addition of positive and negative numbers. Start with the one value on the number line, then add the second value. If the second value (that is added) is positive, we move to the right that many spaces.

If the second value (that is added) is negative, we move to the left that many spaces. The value where we land on the number line is the solution for the addition of two integers.

Example 1 : (-4) + (5) = 1
Start at the first number, -4, and travel 5 units to the right.

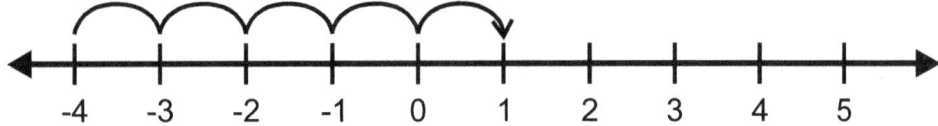

Example 2 : (5) + (-7) = -2
Start at the first number, 5, and travel 7 units to the left.

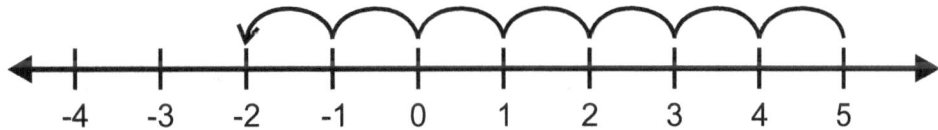

Adding integers using the rules :

Rules for adding integers :

If the signs are the same, add their absolute values, and keep the common sign.

If the signs are different, find the difference between the absolute values of the two numbers, and keep the sign of the number with the greater numerical value.

To the Tune of "Row Your Boat"

Same signs add and keep
Different signs subtract
Keep the sign of the greater digits
then you'll be exact

INTEGERS

Notes

Subtacting integers using a number line :

A number line is helpful in understanding subtraction of positive and negative values. Start with the first value on the number line, then subtract the second value. If the second value (that is subtracted) is positive, we move to the left that many spaces.

If the second value (that is subtracted) is negative, we move to the right that many spaces. This is because subtraction a negative is the same as adding.
The value where we end on the number line is the answer.

Example 1 : (-2) + (5) = 3
Start at the first number, -2, and travel 5 units to the right.

Subtacting integers using the rules :

Every subtraction problem can be written as an additional problem. When we subtract two integers, just <u>ADD THE OPPOSITE.</u> Subtracting a positive is the same as adding a negative. Subtracting a negative is the same as adding a positive.

Multiplying Integers :

Multiplying integers is same as multiplying whole numbers, except we must keep track of the signs associated to the numbers.

To multiply signed integers, always multiply the absolute values and use these rules to determine the sign of the product value

When we multiply two integers with the same signs, the result is always a positive value.

Positive number X Positive number = Positive number

Negative number X Negative number = Positive number

When we multiply two integers with different signs, the result is always a negative value.

Positive number X Negative number = Negative number

Negative number X Positive number = Negative number

Positive X Positive :	7 X 6 = 42	negative X negative : -7 X -6 = 42
Positive X negative :	7 X -6 = -42	negative X Positive : -7 X 6 = -42

INTEGERS

Dividing Integers :

Division of integers is similar to the division of whole numbers, except we must keep track of the signs associated.

To divide signed integers, we must always divide the absolute values and use the below rules to find the quotient value.

When we divide two integers with the same signs, the result is always a positive value.

$$\text{Positive} \div \text{Positive} = \text{Positive}$$

$$\text{Negative} \div \text{Negative} = \text{Positive}$$

When we divide two integers with opposite signs, the result is always a negative value.

$$\text{Positive} \div \text{Negative} = \text{Negative}$$

$$\text{Negative} \div \text{Positive} = \text{Negative}$$

Examples :

Positive ÷ Positive :   81 ÷ 9 = 9   Positive   negative :   81 ÷ -9 = -9

negative ÷ negative :   -81 ÷ -9 = 9   negative   Positive :   -81 ÷ 9 = -9

Golden Rules of Integers :

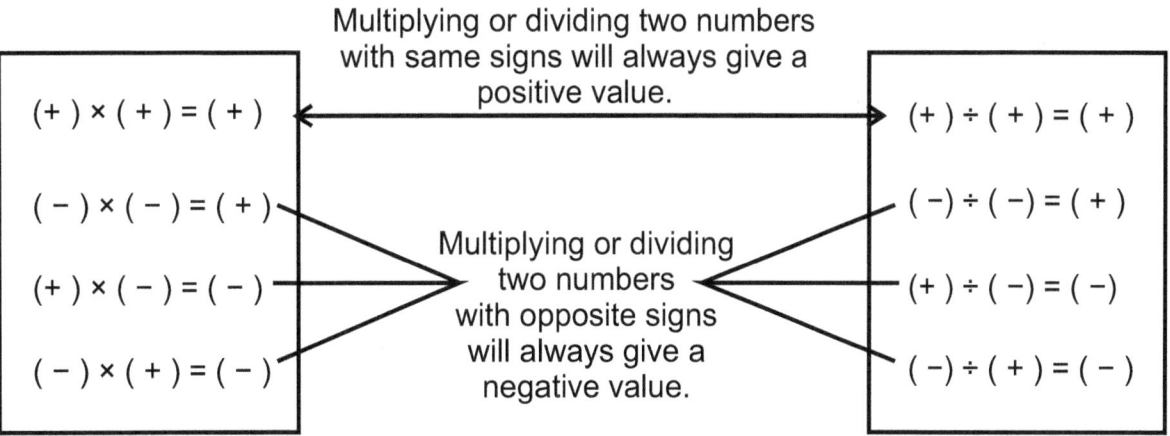

Factors and Exponents

Introduction

Basic rules of number system involves addition, subtraction, multiplication and division. As the topic of "Numbers" involves some more useful concepts like LCM and GCD, we shall study them in this chapter. For finding LCM and GCD, the divisibility rules are useful in one way or the other. Hence let us start the learning the divisibility rules.

Divisibility:

In general, if two natural numbers a and b are such that, when 'a' is divided by 'b', a remainder of zero is obtained, we say that 'a' is divisible by 'b'.

For example, 12 is divisible by 3 because 12 when divided by 3, the remainder is zero.

Also, we say that 12 is not divisible by 5, because 12 when divided by 5, it leaves a remainder 2.

Tests of Divisibility:

We now study the methods to test the divisibility of natural numbers with 2, 3, 4, 5, 6, 8, 9 and 11 without performing actual division.

Test of Divisibility by 2:

A natural number is divisible by 2, if its units digit is divisible by 2, i.e., the units place is either 0 or 2 or 4 or 6 or 8.

Examples : The numbers 4096, 23548 and 34052 are divisible by '2' as they end with 6, 8 and 2 respectively.

Test of Divisibility by 3:

A natural is divisible by 3 if the sum of its digits is divisible by 3.

Example: Consider the number 2143251.

The sum of the digits of 2143251 (2 + 1 + 4 + 3 + 2 + 5 + 1) is 18.

As 18 is divisible by 3, the number 2143251, is divisible by 3.

Factors and Exponents

Test of Divisibility by 4:

A natural number is divisible by 4, if the number formed by its last two digits is divisible by 4.

Examples: 4096, 53216, 548 and 4000 are all divisible by 4 as the numbers formed by taking the last two digits in each case is divisible by 4.

Test of Divisibility by 5:

A natural number is divisible by 5, if its units digit is either 0 or 5.

Examples: The numbers 4095 and 235060 are divisible by 5 as they have in their units place 5 and 0 respectively.

Test of Divisibility by 6:

A number is divisible by 6, if it is divisible by both 2 and 3.

Examples: Consider the number 753618

Since its units digit is 8, so it is divisible by 2. Also its sum of digits = 7 + 5 + 3 + 6 + 1 + 8 = 30, As 30 is divisible by 3, so 753618 is divisible by 3.
Hence 753618 is divisible by 6.

Test of Divisibility by 8:

A number is divisible by 8, if the number formed by its last three digits is divisible by 8.

Examples: 15840, 5432 and 7096 are all divisible by 8 as the numbers formed by last three digits in each case is divisible by 8.

Test of Divisibility by 9:

A natural number is divisible by 9, if the sum of its digits is divisible by 9.

Examples:

(i) Consider the number 125847.
Sum of digits = 1 + 2 + 5 + 8 + 4 + 7 = 27. As 27 is divisible by 9, the number 125847 is divisible by 9.

(ii) Consider the number 145862.
Sum of digits = 1 + 4 + 5 + 8 + 6 + 2 = 26. As 26 not divisible by 9, the number 145862 is not divisible by 9.

Factors and Exponents

Test of Divisibility by 11:

A number is divisible by 11, if the difference between the sum of the digits in odd places and sum of the digits in even places of the number is either 0 or a multiple of 11.

Examples:
(i) Consider the number 9582540
Now (sum of digits in odd places) - (sum of digits in even places)
= (9 + 8 + 5 + 0) - (5 + 2 + 4)
= 11, which is divisible by 11.
Hence 958254 is divisible by 11.

(ii) Consider the number 1453625
Now, (sum of digits at odd places) - (sum of digits at even places)
= (1 + 5 + 6 + 5) - (4 + 3 + 2)
= 17 - 9 = 8, which is not divisible by 11

Some Additional Results:

If a natural number N is divisible by two natural numbers a and b, then N is divisible by the product of a and b, if and only if a and b are co-primes.

Examples:
(i) 345 is divisible by 3 as well as by 5, as 3 and 5 are co-primes, 345 is divisible by 15.
(ii) 120 is divisible by 8 and 10, but it is not divisible by 80.

Factors and Multiples:

Lets learn the concepts of factors and multiples.

Definition:

If 'b' divides 'a' leaving a zero remainder, then 'b' is called a factor or divisor of 'a' and 'a' is called the multiple of 'b'.
For example, 6 = 2 3
Here 2 and 3 are factors of 6 (or) 2 and 3 are divisors of 6.
And, 6 is a multiple of 3, 6 is a multiple of 2.

Examples: (i) The factors of 24 = {1, 2, 3, 4, 6, 8, 12, 24}
(ii) The factors of 256 = {1, 2, 4, 8, 16, 32, 64, 128, 256}

Observations:

One is the factor of every natural number and it is the least of the factors of any natural number.
Every natural number is the factor of itself and it is the greatest.

Some Additional Results:

Unique Prime Factorisation Theorem: "Any natural number greater than 1 can be divided into a prime number or a composite number."

For example:

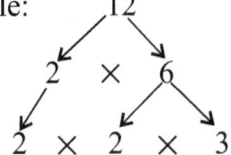

6 is a composite number

All factors are prime numbers.

If a is a composite number of the form $a = b^p c^q d^r \ldots$ where b, c, d ... are district prime factors, then the number of factors of $a = (p+1)(q+1)(r+1)\ldots$

If a is a composite number of the form $a = b^p c^q d^r \ldots$ where b, c, d .... are district prime factors, then the sum of all the factors of $a = \frac{(b^{p-1} - 1)}{(b-1)} \times \frac{(c^{q-1} - 1)}{(c-1)} \times \frac{(d^{r-1} - 1)}{(d-1)} \ldots$

Perfect Numbers:

A number for which sum of all its factors is twice the number itself is called a perfect number.

Observation:

Euler proved that if $2^k - 1$ is a prime number, then $2^{k-1}(2^k - 1)$ is a perfect number.
A perfect number can never be a prime number.

Examples:
(i) Consider the composite number 6.
Factors of 6 = {1, 2, 3, 6}
Sum of factors = (1 + 2 + 3 + 6) = 12
Clearly, the sum of the factors of 6 is twice the number itself.
Hence 6 is a perfect number.

Example:
(ii) Consider the composite number 48
Factors of 48 = {1, 2, 3, 4, 6, 8, 12, 16, 24, 48}
Sum of factors = 1 + 2 + 3 + 4 + 6 + 8 + 12 + 16 + 24 + 48
= 124 2 48
Clearly, 48 is not a perfect number.

Factors and Exponents

Notes

Greatest Common Divisor [GCD] (or) Greatest Common Factor [GCF] (or) Highest Common Factor [HCF]

Definition :

"The greatest common factor of two or more natural numbers is the largest factor in the set of common factors of those numbers." In other words, the GCD (or) GCF of two or more numbers is the largest number that divides each of them exactly.

Example : Find the GCF of 72 and 60.

Solution : Let the set of factors of 72 be A.
A = {1, 2, 3, 4, 6, 8, 9, 12, 18, 24, 36, 72}
Let the set of factors of 60 be B.
B = {1, 2, 3, 4, 5, 6, 10, 12, 15, 20, 30, 60}
The set of common factors for 72 and 60 is A B = {1, 2, 3, 4, 6, 12}
The greatest element in this set is 12
The GCF (or) GCD for 72 and 60 is 2.

Observations :
If two numbers have no factors in common, then their GCF is unity.
i.e., GCF of prime numbers and co-prime numbers is unity.

Methods of finding GCF :

Factors Method :

When the numbers whose GCF has to be found are relatively small, this is the best suited method. Here we resolve the given numbers into their prime factors and find out the largest factor in the set of common factors to given numbers.

This method can be easily applied to any number of numbers.

Examples : (i) Find the GCF of 24 and 36.

Solution :

Resolving given numbers into product of prime factors, we have
36 = ② × ② × ③ × 3
24 = 2 × ② × ② × ③
The common factors to both the numbers are circled.
Now GCF = product common factors of given numbers = 2 × 2 × 3 = 12
GCF (24, 36) = 12

Factors and Exponents

Notes

(ii) Find the GCF of 12, 18 and 24.

Solution :

Resolving given numbers into product of prime factors;
12 = ② × 2 × ③
18 = ② × 3 × ③
24 = ② × 2 × 2 × ③

GCF = product of common factors of 12, 18 and 24
     = 2 × 3 = 6
GCF = 6

Division Method :

When the numbers whose GCF has to be found are very large, it is time consuming to write down sets of common factors to given numbers;

In this case, we use the method of Long Division. This method was proposed by Euclid and the following steps are involved in it.

Step 1 : Divide the larger number by the smaller number. If the remainder is zero, the divisor is the GCF, otherwise not.

Step 2 : Let the divisor in step 2 be the dividend now, and the remainder of step 1 become the divisor of step 2. Again, if the remainder is zero, the divisor is GCD. Otherwise, step 2 has to be repeated.

Example : Find the GCF of 64 and 56

Solution :
64 divided by 56, quotient is 1 and remainder is 8. Because the reminder is not zero; 56 is not the GCD.

Proceeding further, as mentioned in step 2, 56 is dividend and 8 is divisor. The quotient is 3 and remainder is zero.

Because the remainder is zero, the divisor 8 is the GCD.

GCF of three numbers using division method:

The GCF of 3 numbers is found out by finding GCF of any 2 numbers and GCF of the remaining number with the GCF obtained above.
i.e., GCF (a, b, c) = GCF [GCF (a, b), c].

Factors and Exponents

This process can be extended to any number of numbers.

Example: Find the GCF of 25, 45 and 75.

Solution: Let us first find the GCF of 25 and 45

$$
\begin{array}{r}
25\overline{)45}(1 \\
\underline{25} \\
20\overline{)25}(1 \\
\underline{20} \\
5\overline{)20}(4 \\
\underline{20} \\
0
\end{array}
$$

GCF (25, 45) = 5
The GCF of (5, 75) is 5.
The GCF of 25, 45 and 75 is 5.

Observations:

The process of dividing the divisor with quotient is to be repeated until remainder as zero is obtained

If the zero remainder is obtained when the divisor is 1, then the GCD is '1'.

GCD is '1', means that two numbers are relatively prime or co-prime; because they do not have any factor in common other than 1. For eg; 12 and 13 are co-primes.

Some Additional Results:

The largest number which divides p, q and r to give remainders of s, t and u respectively will be the GCD of the three numbers (p − s), (q − t) and (r − u).

The largest number which divides the numbers p, q and r and gives the same remainder in each case will be the GCD of the differences of two or the three numbers (p − q, q − r, p − r).

Least Common Multiple [LCM]:

Definition:

"The least common multiple of two or more natural numbers is the least of their common multiples". In other words, the LCM of two or more numbers is the least number which can be divided exactly by each of the given numbers.

Factors and Exponents

Note : If the set of common multiples is denoted by C, then N and the number of elements in C is infinite and the least element in C is their LCM.

Example : Find L.C.M. of 24 and 36.

Solution : Resolving 24 and 36 into product of prime factors
24 = ②×②× 2 ×③
36 = ②×②×③× 3

The common prime factors of 24 and 36 are 2, 2 and 3. (which are circled)

The remaining prime factors of 24 is 2. (which is not circled).

The remaining prime factors of 36 is 3. (which is not circled).

LCM = Common factors of 24 the prime factors left in 24 the prime factors left in 36
= 2 × 2 × 3 × 2 × 3
= 72

Methods of finding LCM :

Factors Method :

Here the given numbers are decomposed into product of prime factors; from which, the least common multiple is found by multiplying the terms containing factors of numbers raised to their highest powers.

Example : Find the LCM of 32 and 24.

Solution : Resolving given numbers into product of common factors, we have
$32 = 2^5 ; 24 = 2^3 \times 3$
LCM = product of terms containing highest powers of factors 2, 3
$= 2^5 \cdot 3 = 96$

LCM of three numbers using factors method:

The method above can be extended in a similar way to three numbers. This is illustrated below:

Example : Find the LCM of 12, 48 and 36

Solution : Resolving given numbers into product of common factors, we have
$12 = 2^2 \times 3^1 ; 48 = 2^4 \times 3^1 ; 36 = 2^2 \times 3^2$. Then their LCM $= 2^2 \times 3^2 = 16 \times 9 = 144$

Factors and Exponents

Notes

Synthetic Division Method of Finding LCM:

LCM of numbers can also be found using synthetic division method. This is illustrated below:

Example: Find the LCM of 144 and 156.

Solution: Using synthetic division, we have:

```
2 | 144, 156
2 |  72,  78
3 |  36,  39
  |  12,  13
```

LCM = 2 . 2 . 3 . 12 . 13 = 1872

Examples: Find the LCM of 12, 18 and 24

Solution: Using synthetic division;

```
2   12, 18, 24
2    6,  9, 12
2    3,  9,  6
     1,  3,  2
```

LCM = 2 . 2 . 2 . 1 3 . 2 = 48

Factors and Exponents

Relationship between LCM and GCF:

The LCM and GCF of two given numbers are related to the given numbers by the following relationship.
Product of the numbers = LCM × GCF

where, LCM denotes the LCM of the given numbers and GCF denotes the GCF of the given numbers.

Example: Consider two numbers, 24 and 36.
These can be resolved into product of prime factors as below:
$24 = 2^3 \times 3$
$36 = 2^2 \times 3^2$
Now LCM $(24, 36) = 2^3 \times 3^2 = 72$
GCF $(24, 36) = 2^2 \times 3 = 12$

Now; Product of numbers = $24 \times 36 = 2^5 \times 3^3 = 864$
Product of LCM and GCF = $72 \times 12 = 2^5 \times 3^3 = 864$

Clearly, Product of the numbers = Product of the LCM and GCF.

Relatively Prime Numbers:

Definition:

If two numbers do not have any common factors other than 1, then they are called relatively prime numbers or co-prime numbers.

Concept: We know, every number has at least two factors, 1 and itself. If it has more than two factors, it is a composite number and if it does not have any factor except 1 and itself, it is a prime number.

But if two numbers (prime or composite) are such that they have only one common factor '1' are called relatively prime.

Example: Consider three numbers 8, 18 and 25
Now,
A, The set of factors of 8 = {1, 2, 4, 8}
B, The set of factors of 18 = {1, 2, 3, 6, 9, 18}
C, The set of factors of 25 = {1, 5, 25}

Now, A B = {1, 2} ; B C = {1} ; C A = {1}

Clearly the common factors for both (18, 25) and (8, 25) is 1 only.

Factors and Exponents

They are generally written as (18, 25) = 1 and (8, 25) = 1

Note : Here; 18, 25 and 8 are not prime numbers (composite) but they are relatively prime numbers.

Observations :

The G.C.D. of two relatively prime numbers is 1 and their LCM is product of the numbers

Any two prime numbers are always relatively prime to each other.

Two relatively prime numbers need not be prime numbers.

LCM and GCF of Fractions :

The LCM and GCF of fractions can be determined by the following relations :

$$\text{LCM of fractions} = \frac{\text{LCM of numerators}}{\text{GCD of denominators}}$$

$$\text{GCF of fractions} = \frac{\text{GCD of numerators}}{\text{LCM of denominators}}$$

Examples : Find the GCD and LCM of $\frac{4}{5}$, $\frac{2}{5}$ and $\frac{3}{4}$

Solutions : $\text{LCM}\left(\frac{4}{5}, \frac{2}{5}, \frac{3}{4}\right) = \frac{\text{LCM}(4, 2, 3)}{\text{HCF}(5, 5, 4)} = \frac{4 \times 3}{5}$

$\text{GCD}\left(\frac{4}{5}, \frac{2}{5}, \frac{3}{4}\right) = \frac{\text{HCF}(4, 2, 3)}{\text{LCM}(5, 5, 4)} = \frac{1}{5 \times 4} = \frac{1}{20}$

Factors and Exponents

Introduction of Exponents

Whenever any integer, let us say x, is added n times, the result obtained will be equal to n times x i.e. nx. But in case, if the integer x is multiplied for n times, the result obtained will be equal to x^n (which is called as exponential form). The problems relating to these will be studied under 'Exponents'. We shall look at the rules/properties pertaining to these exponential numbers in this chapter.

Rational Exponents and Radicals

If 'a' is any real number and 'n' is a positive integer, then the product $a \times a \times a \times \text{----} \, n$ times is represented by the notation a^n. This notation is referred to as exponential form. In the above notation, a is called the base and n is called the power or exponent or index (plural of index is indices). a^n is read as 'n^{th} power of a' or 'a to the power n'.

Example: $6 \times 6 \times 6 \times 6 \times 6 \times 6 \times 6$ can be written as 6^7. Here 6 is called as base and 7 is called as index (or exponent).

For a non-zero rational number 'a' with a negative integral exponent 'm' the following result can be observed.

$a^m = a^{-n} = a^{-1} \times a^{-1} \times a^{-1} \times a^{-1} \times \text{----} \times \text{---} \, n$ times

$= \dfrac{1}{a} \times \dfrac{1}{a} \times \dfrac{1}{a} \times \dfrac{1}{a} \text{----} \, n \text{ times} = \left(\dfrac{1}{a}\right)^n = \dfrac{1}{a^n}$

Example: $6^{-3} = \left(\dfrac{1}{6}\right)^3$

Rational Indices:

1. n^{th} root of a :
   A real number x is said to be the nth root of a if $x^n = a$; where a is any real number and n is a positive integer.

Parts of the exponent:

This is read as "Seven to the fourth power"

Factors and Exponents

Notes

The nth root of a can be represented as $a^{1/n}$ or $\sqrt[n]{a}$. Here $a^{1/n}$ is called exponential form and the form $\sqrt[n]{a}$ is called radical form. The sign $\sqrt[n]{}$ is called radical sign and $\sqrt[n]{a}$ is called radical. The number n is a positive integer is called the index of radical and a is called the radicand.

Example: We know that $32 = 2^5$. So we can say that 2 is 5th root of 32. It is written as $32^{1/5} = 2$ or $\sqrt[5]{32} = 2$
Similarly $\sqrt[3]{64} = 3$, $\sqrt[4]{625} = 5$, $\sqrt[6]{64} = 2$, etc.

Note:

(i) If n is negative as in case $64^{-\frac{1}{3}}$, we write the radical form as follows;

$$64^{-\frac{1}{3}} = \left(\frac{1}{64}\right)^{\frac{1}{3}} = \sqrt[3]{\frac{1}{64}}.$$

The radical form of $64^{-\frac{1}{3}}$ should not be taken as $\sqrt[-3]{64}$ as in the radical form $\sqrt[n]{a}$ of n is a positive integer.

i.e., $a^{-1/n} = \sqrt[n]{\frac{1}{a}}$, where n is a positive integer.

(ii) $\sqrt[n]{a}$ is positive for $a > 0$ and n being a positive integer.
Example: $\sqrt[5]{32} = 2$, $\sqrt[6]{64} = 2$, $\sqrt[3]{27} = 3$, etc.

(iii) $\sqrt[n]{a}$ is negative for $a < 0$ and n being any odd positive integer.
Example: $\sqrt[3]{-8} = -2$, $\sqrt[7]{-128} = -2$, $\sqrt[5]{-243} = -3$, etc.

(iv) $\sqrt[n]{a}$ does not exist in set of real numbers, for $a < 0$ and n being even positive integer.
Example: $\sqrt[2]{-16}$, $\sqrt[4]{-256}$, $\sqrt[6]{-64}$ etc doesn't exist.

Each positive number has two square roots, one positive and the other negative.
Example: $\sqrt{36} = 6$ or -6 (since $6^2 = (-6)^2 = 36$).

Factors and Exponents

Notes

If a is a positive rational number and n = p/q is a positive rational exponent, then we can define $a^{p/q}$ in two ways.

(1) $a^{\frac{p}{q}}$ is the qth root of a^p, i.e. $a^{\frac{p}{q}} = (a^p)^{\frac{1}{q}}$

(2) $a^{\frac{p}{q}}$ is the pth power of qth root of a, i.e. $a^{\frac{p}{q}} = \left(a^{\frac{1}{q}}\right)^p$.

Laws of indices:

For all real numbers a and b and all rational numbers m and n, we have

(i) $a^m \times a^n = a^{m+n}$

Examples: (1) $2^3 \times 2^6 = 2^{3+6} = 2^9$

(2) $\left(\frac{5}{6}\right)^4 \times \left(\frac{5}{6}\right)^5 = \left(\frac{5}{6}\right)^{4+5} = \left(\frac{5}{6}\right)^9$

(3) $5^{2/3} \times 5^{4/3} = 5^{(2/3 + 4/3)} = 5^{6/3} = 5^2$

(4) $2^3 \times 2^4 \times 2^5 \times 2^8 = 2^{(3+4+5+8)} = 2^{20}$.

(5) $(\sqrt{7})^3 \times (\sqrt{7})^{\frac{5}{2}} = (\sqrt{7})^{3+\frac{5}{2}} = (\sqrt{7})^{\frac{11}{2}}$

(ii) $a^m \div a^n = a^{m-n}$, $a \neq 0$

Examples: (a) $7^8 \div 7^3 = 7^{8-3} = 7^5$

(b) $\left(\frac{7}{3}\right)^9 \div \left(\frac{7}{3}\right)^5 = \left(\frac{7}{3}\right)^{9-5} = \left(\frac{7}{3}\right)^4$

(c) $9^{\frac{2}{3}} \div 9^{\frac{1}{6}} = 9^{\left(\frac{2}{3}-\frac{1}{6}\right)} = 9^{\left(\frac{4-1}{6}\right)} = 9^{\frac{3}{6}} = 9^{\frac{1}{2}}$

(d) $\left(\frac{5}{7}\right)^{\frac{8}{9}} \div \left(\frac{5}{7}\right)^{\frac{1}{3}} = \left(\frac{5}{7}\right)^{\left(\frac{8}{9}-\frac{1}{3}\right)} = \left(\frac{5}{7}\right)^{\frac{8-3}{9}} = \left(\frac{5}{7}\right)^{\frac{5}{9}}$

Factors and Exponents

Note: $a^n \div a^n = 1$
or $a^{n-n} = a^0 = 1$
$\therefore a^0 = 1, a \neq 0$

(iii) $(a^m)^n = a^{m \times n}$

Examples: (a) $(5^2)^3 = 5^{2 \times 3} = 5^6$

(b) $\left[\left(\dfrac{2}{3}\right)^4\right]^5 = \left(\dfrac{2}{3}\right)^{4 \times 5} = \left(\dfrac{2}{3}\right)^{20}$

(c) $\left[\left(\dfrac{5}{7}\right)^{\frac{2}{3}}\right]^{\frac{9}{8}} = \left(\dfrac{5}{7}\right)^{\left(\frac{2}{3} \times \frac{9}{8}\right)} = \left(\dfrac{5}{7}\right)^{\frac{3}{4}}$

(iv) $\left(\dfrac{a}{b}\right)^n = \dfrac{a^n}{b^n}$

Example: $\left(\dfrac{4}{5}\right)^7 = \dfrac{4^7}{5^7}$

Note: Conversely we can write $\left(\dfrac{a^n}{b^n}\right) = \left(\dfrac{a}{b}\right)^n$

Example: $\dfrac{8}{27} = \dfrac{2^3}{3^3} = \left(\dfrac{2}{3}\right)^3$

(v) $(ab)^n = a^n \times b^n$

Examples: (a) $20)^5 = (4 \times 5)^5 = 4^5 \times 5^5$
(b) $(42)^7 = (2 \times 3 \times 7)^7 = 2^7 \times 3^7 \times 7^7$

Note: Conversely we can write $a^n \times b^n = (ab)^n$

Factors and Exponents

Examples: (a) $4^8 \times 5^8 = (4 \times 5)^8 = 20^8$

(b) $\left(\dfrac{2}{3}\right)^5 \times \left(\dfrac{9}{8}\right)^5 = \left(\dfrac{2}{3} \times \dfrac{9}{8}\right)^5 = \left(\dfrac{3}{4}\right)^5$

(vi) $a^{-n} = \dfrac{1}{a^n}$, $a \neq 0$

Example: $2^{-4} = \dfrac{1}{2^4}$, $5^{-1} = \dfrac{1}{5}$

Note: $a^{-1} = \dfrac{1}{a^1} = \dfrac{1}{a}$

(vii) $\left(\dfrac{a}{b}\right)^n = \left(\dfrac{b}{a}\right)^n$

Examples: (a) $\left(\dfrac{5}{9}\right)^3 = \left(\dfrac{9}{5}\right)^{-3}$

(b) $\left(\dfrac{1}{5}\right)^{-1} = \left(\dfrac{5}{1}\right)^1 = 5$

Note: $\left(\dfrac{1}{a}\right)^{-1} = \left(\dfrac{a}{1}\right)^1 = a$

(viii) If $a^m = a^n$, then $m = n$, where $a \neq 0$, $a \neq 1$

Examples: (a) If $5^p = 5^3 \Rightarrow p = 3$

(b) If $4^p = 256$
$4^p = 4^4 \Rightarrow p = 4$

(ix) For positive numbers a and b, if $a^n = b^n$, $n \neq 0$, then $a = b$ (when n is odd)

Examples: (a) If $5^7 = p^7$, then clearly $p = 5$.

(b) If $(5)^{2n-1} = (3 \times p)^{2n-1}$, then clearly $5 = 3p$ or $p = 5/3$

Factors and Exponents

(x) If $p^m \times q^n \times r^s = p^a q^b r^c$, then $m = a$, $n = b$, $s = c$, where p, q, r are different primes.

Examples: (a) If $40500 = 2^a \times 5^b \times 3^c$, then find $a^a \times b^b \times c^c$

$$\begin{array}{r|l} 2 & 40,500 \\ 2 & 20,250 \\ 5 & 10,125 \\ 5 & 2,025 \\ 5 & 405 \\ 3 & 81 \\ 3 & 27 \\ 3 & 9 \\ & 3 \end{array}$$

$\therefore 40500 = 2^2 \times 5^3 \times 3^4 = 2^a \times 5^b \times 3^c$

$\therefore a = 2, b = 3, c = 4$, [Using the above law].

$\therefore a^a \times b^b \times c^c = 2^2 \times 3^3 \times 4^4 = 27,648$

Example 8: (a) $20)^5 = (4 \times 5)^5 = 4^5 \times 5^5$

(b) $(42)^7 = (2 \times 3 \times 7)^7 = 2^7 \times 3^7 \times 7^7$

Note: Conversely we can write $a^n \times b^n = (ab)^n$

Factors and Exponents

Example 9 :

(a) $\left((5)^3\right)^2 = 5^{3 \times 2} = (5)^6 = 5 \times 5 \times 5 \times 5 \times 5 \times 5 = 15625$

(b) $(2)^3 \times (2)^5 = (2)^{3+5} = (2)^8 = 2 \times 2 \times 2 \times 2 \times 2 \times 2 \times 2 \times 2 = 256$

(c) $(7)^0 = 1$ $\boxed{\text{Any base value rise to the power zero is always equal to 1}}$

(d) $(3)^{-4} = \dfrac{1}{(3)^4} = \dfrac{1}{3 \times 3 \times 3 \times 3} = \dfrac{1}{81}$

(d) $\dfrac{(8)^7}{(8)^5} = (8)^{7-5} = (8)^2 = 64$

(e) $\dfrac{(9)^4}{(9)^7} = (9)^{4-7} = (9)^{-3} = \dfrac{1}{(9)^3} = \dfrac{1}{9 \times 9 \times 9} = \dfrac{1}{729}$

(f) $(2)^4 = 2 \times 2 \times 2 \times 2 = 16$ 　　(g) $(-2)^4 = -2 \times -2 \times -2 \times -2 = 16$

(h) $-(2)^4 = -(2 \times 2 \times 2 \times 2) = -16$ 　　(i) $-(2)^3 = -(2 \times 2 \times 2) = -8$

(j) $(-2)^3 = -2 \times -2 \times -2 = -8$ 　　(k) $-(-2)^3 = -(-2 \times -2 \times -2) = -(-8) = 8$

Tip 1 : When the exponent is an even number the simplified value is always positive, when the base has a positive or negative value.

Tip 2 : When the exponent is an odd number the simplified value is always positive, when the base has a positive value.

Tip 3 : When the exponent is an odd number the simplified value is always negative, when the base has a negative value.

PERCENTAGES

The percentage symbol is a representation of percentage. In statistics, percentages are often left in their base form of 0 - 1, where 1 represents the whole. We multiply the decimal by a factor of 100 to find the percentage.

Percent Equation form :

Percentages can be setup as proportions. The parts of a percent can be found by setting up a proportion as below.

$$\frac{is}{of} = \frac{percent}{100} \quad OR \quad \frac{part}{total} = \frac{percent}{100}$$

Before we calculate a percentage, we should understand exactly what a percentage is ? The word percentage comes from the word percent. If you split the word percent into its root words, you see "per" and "cent." Cent is an old European word with French, Latin, and Italian origins meaning "hundred". So, percent is translated directly to "per hundred." If we have 39 percent, we literally have 39 per 100. If it snowed 16 times in the last 100 days, it snowed 16 percent of the time.

Whole numbers, decimals and fractions are converted to percentages. Decimal format is easier to convert into a percentage. Fractions can be converted into decimals and then to percentages.

Converting a decimal to a percentage is as simple as multiplying it by 100. To convert 0.92 to a percent, simply multiple 0.92 by 100.

0.92 × 100 = 92%

A percentage is an expression of part of the whole. Nothing is represented by 0%, and the whole amount is 100%. Everything else is somewhere in between 0 and 100.

For example, say you have 20 muffins. If you share 12 muffins with your friends, then you have shared 12 out of the 20 muffins ($\frac{12}{20}$ × 100% = 60% shared). If 20 muffins are 100% and you share 60% of them, then 100% - 60% = 40% of the muffins are still left.

PERCENTAGES

Notes

Determine the part of the whole :

Given the value for part of the whole and the whole.
              OR
Given two parts that make up the whole.

It is important to differentiate what the percentage is "of."
Example : A jar contains 99 gold beads and 51 blue beads. A total of 150 beads are in the jar. In this example, 150 beads makes up a whole jar of beads, i.e. 100%.

1. **Put the two values into a fraction.** The part goes on top of the fraction (numerator), and the whole goes on the bottom (denominator). Therefore the fraction in this case is $\frac{99}{150}$ (part/whole) or $\frac{51}{150}$ (part/whole).

2. **Convert the fraction into a decimal.** Convert $\frac{99}{150}$ into decimal, divide 99 by 150

   $\frac{99}{150} = 0.66$

Converting the decimal into a percent :

Multiply the result obtained in the step above by 100% (per 100 = *per cent*).
0.66 multiplied by 100 equals 66%. Gold beads are 66% of the total beads in the jar.

Converting the percentage into a decimal : Working backward from before, divide the percentage by 100, or you can multiply by 0.01.

66% = $\frac{66}{100}$ = 0.66

Re-word the problem with your new values :

Given in the form of "**X** of **Y** is **Z**." X is the decimal form of your percent, "of" means to multiply, Y is the whole amount, and Z is the answer.

Example: 3% of $25 is 0.75.
           The amount of interest accrued each day on a 3% of loan amount $25
           s 0.75

PERCENTAGES

Discounts :

Discounts are offered on the original price of the item on sale.

1. The discount percent need to be subtracted from the whole, meaning if a discount of 20% is offered on an item priced at $60 then sale price is 80% of $60.
   100% - 20% = 80% (whole percent - discount percent).

   $80\% \text{ of } \$60 = \dfrac{80}{100} \times 60 = \$48$

Markup percentage :

Markup percentage is also referred as gain percentage. Markups are added to the original price of the item before selling it. The greater the markup percentage higher are the profits.

1. The markup percent needs to be added to the whole, meaning if a markup of 20% is added on an item priced at $60 then sale price is 120% of $60.
   100% + 20% = 120% (whole percent + markup percent).

   $120\% \text{ of } \$60 = \dfrac{120}{100} \times 60 = \$72$

Tax percentage :

Taxes are added to the original price of the item before selling it to the customer.

1. The tax percent needs to be added to the whole, meaning if a tax of 5% is added on an item priced at $60 then sale price is 105% of $60.
   100% + 5% = 105% (whole percent + tax percent).

   $105\% \text{ of } \$60 = \dfrac{105}{100} \times 60 = \$63$

Simple Interest :

Simple interest is an easy way of calculating interest on a loan amount.

$$I = PTR$$

I = Simple Interest
P = Principle amount of loan
T = Term of the loan in years
R = Rate of interest in decimal form.

Remember final loan amount (A) = P + I

PERCENTAGES

Or

$$I = \frac{PTR}{100}$$

I = Simple Interest
P = Principle amount of loan
T = Term of the loan in years
R = Rate of interest in numerical value only (not in decimal form since the formula has 100 in the denominator).

Remember final loan amount (A) = P + I

Example : Dan took a loan of $1,650 at 7% rate of interest for 5 years from his friend Johnson. Using simple interest find the loan amount to be paid to Johnson at the end of the term.

$$I = \frac{PTR}{100}$$

P = $1,650
R = 7
T = 5 years

$$I = \frac{1650 \times 7 \times 5}{100}$$

$$I = \frac{57750}{100}$$

$$= \$577.50$$

Interest to be paid at the end of the loan term is $577.50
Remember the original loan amount $1,650 also need to be paid of.
Amount to be paid of (A) = P + I

$$A = \$1,650 + \$577.50$$
$$= \$2,227.50$$

 **PERCENTAGES**

Notes

Compound interest :

Compound interest is a complex way of calculating interest on a loan amount and the interest on the interest accurred.

$$A = P\left(1 + \frac{r}{n}\right)^{nt}$$

Where A = Amount at the end of the loan term
Note :It includes principle amount and the interest

    P = Principle loan amount
    r = Rate of loan or loan percent or interest rate.
Note : Rate needs to be converted to decimal.

    n = Number of times interest is compounded per year.
    t = Loan term in years.

Remember final loan amount (A) = P + I

OR

$$A = P\left(1 + \frac{r}{100n}\right)^{nt}$$

Where A = Amount at the end of the loan term
Note :It includes principle amount and the interest

    P = Principle loan amount
    r = Rate of loan or loan percent or interest rate.
Note : Rate need not be converted to decimal.

    n = Number of times interest is compounded per year.
    t = Loan term in years.

Remember final loan amount (A) = P + I

PERCENTAGES

Example : Dan took a loan of $49,100 at 14% rate of interest for 2 years compounded annually from his friend Johnson. Using compound interest find the loan amount to be paid to Johnson at the end of the term.

$$A = P\left(1 + \frac{r}{100n}\right)^{nt}$$

P = $49,100
R = 14
T = 2 years
n = 1

$$A = 49100\left(1 + \frac{14}{100}\right)^2$$

$$A = 49100\,(1.14)^2$$

$$A = \$63{,}810.36$$

$$A = P + I \text{ OR } I = A - P$$

Interest I = $63,810.36 - $49,100 = $14,710.36

Interest to be paid at the end of the loan term is $14,710.36
Remember the original loan amount $49,100 also need to be paid of.
Amount to be paid of (A) = P + I

$$A = \$63{,}810.36$$

Dan needs to pay $63,810.36 to Johnson at the end of the loan term.

PERCENTAGES

Example : Dan took a loan of $49,100 at 14% rate of interest for 2 years compounded semi annually from his friend Johnson. Using compound interest find the loan amount to be paid to Johnson at the end of the term.

$$A = P\left(1 + \frac{r}{100n}\right)^{nt}$$

P = $49,100
R = 14
T = 2 years
n = 2

$$A = 49100\left(1 + \frac{14}{200}\right)^4$$

$$A = 49100\,(1.07)^4$$

$$A = \$64{,}360.08$$

$$A = P + I \text{ OR } I = A - P$$

Interest I = $64,360.08 - $49,100 = $15,260.08

Interest to be paid at the end of the loan term is $15,260.08
Remember the original loan amount $49,100 also need to be paid of.
Amount to be paid of (A) = P + I

$$A = \$64{,}360.08$$

Dan needs to pay $64,360.08 to Johnson at the end of the loan term.

Geometry Notes

1. Complementary Angles:
Two angles that add up to 90 degrees are called as Complementary angles.

Example: $\angle X + \angle Y = 90$
$\angle X$ and $\angle Y$ are called complementary angles.

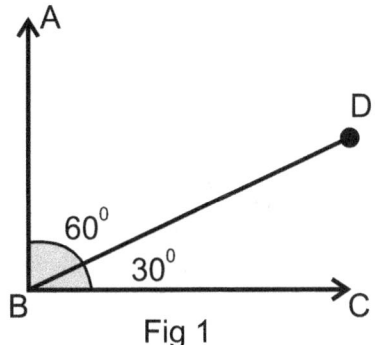

Fig 1

$30^0 + 60^0 = 90^0$
30^0 and 60^0 are complementary angles

2. Supplementary Angles:
Two angles that add up to 180 degrees are called as supplementary angles.
Example: $\angle a + \angle b = 180$
$\angle a$ and $\angle b$ are called complementary angles.

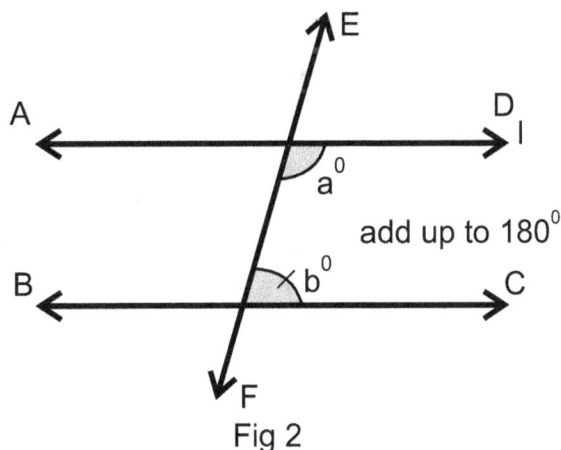

Fig 2

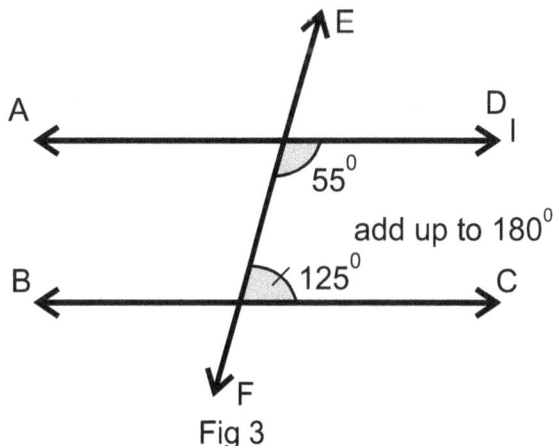

Fig 3

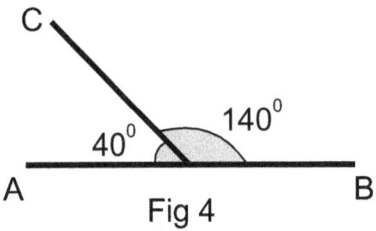

Fig 4

$\angle a + \angle b = 180$; $55^0 + 125^0 = 180^0$
55^0 and 125^0 are called as complementary angles

$\angle a + \angle b = 180$; $40^0 + 140^0 = 180^0$
40^0 and 140^0 are called as complementary angles

GEOMETRY

Notes

3. Vertical Angles:

Vertical angles are pairs of opposite angles made by intersecting lines.
If two angles are vertical, then they are congruent.
Example:
  ∠a and ∠b are called vertical angles
  ∠a = ∠b

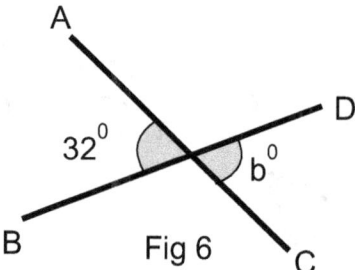

Fig 5   Fig 6

∠a and ∠b are called vertical angles and vertical angles are equal to each other.
In Fig 6 based on the rule of the vertical angles ∠b = 32^0

4. Adjacent Angles:

Two angles that have a common side and a common vertex (corner point), and don't overlap are called adjacent angles

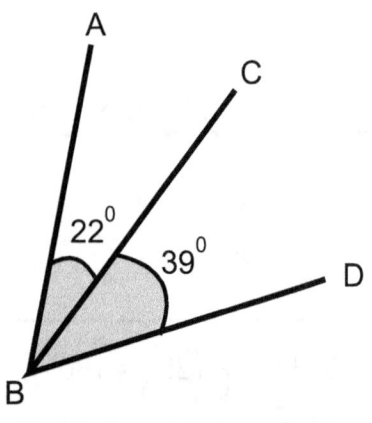

∠ABC and ∠CBD are called as adjacent angles as they share same vertex B

Fig 7

5. Corresponding Angles

When two lines are crossed by another line (Transversal), the angles in matching corners are called as corresponding angles. A pair of angles each of which is on the same side of one of two lines cut by a transversal and on the same side of the transversal

The angles which occupy the same relative position at each intersection where a straight line crosses two others. If the two lines are parallel, the corresponding angles are equal.

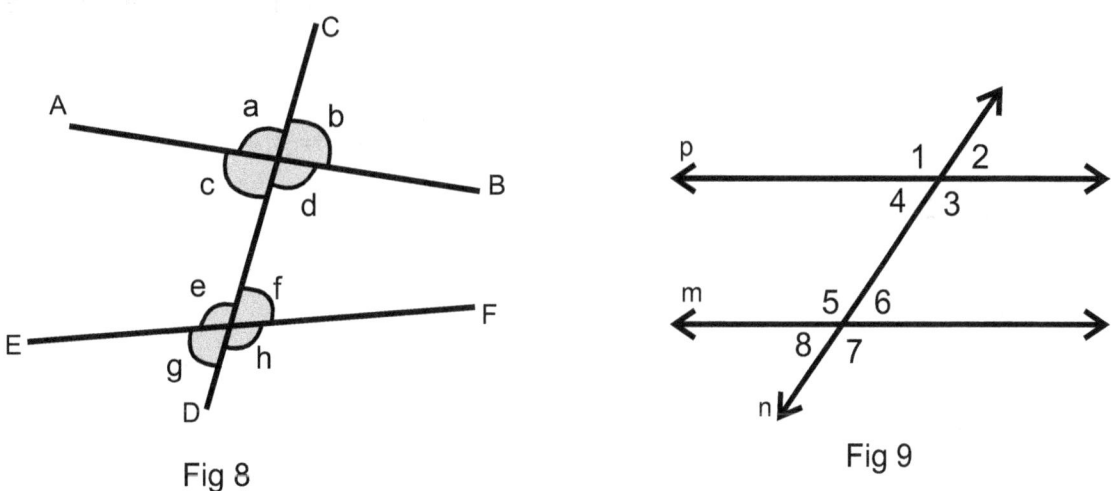

Fig 8   Fig 9

In Fig 8, $\overline{AB}$ and $\overline{EF}$ **are not parallel**, the corresponding angles ∠a and ∠e ; ∠c and ∠g ; ∠b and ∠f ; ∠d and ∠h are **not equal**.

In Fig 9, Lines p and m **are parallel**, the corresponding angles ∠1 and ∠5 ; ∠4 and ∠8 ; ∠2 and ∠6 ; ∠3 and ∠7 **are equal**.

6. Alternate Angles

Two angles, not adjoining one another, that are formed on opposite sides of a line that intersects two other lines. If the original two lines are parallel, the alternate angles are equal.
one of a pair of angles with different vertices and on opposite sides of a transversal at its intersection with two other lines:

1. Alternate Interior Angles are a pair of angles on the inner side (inside) of each of those two intersected lines but on opposite sides of the transversal. If two parallel lines are cut by a transversal, the alternate interior angles are congruent Examples of Alternate Interior Angles In the figure shown, l is the transversal that cut the pair of lines. Angles 3 and 4 and angles 1 and 2 are alternate interior angles.

GEOMETRY

Notes

2: Alternate Exterior Angles are a pair of angles on the outer side (outside) of each of those two intersected lines but on opposite sides of the transversal.

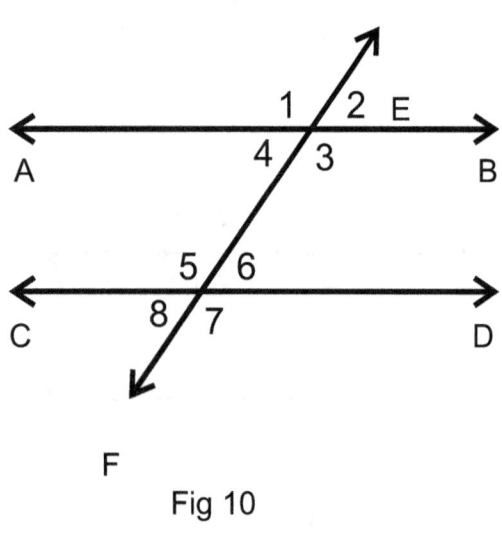

Fig 10

∠3, ∠4, ∠5, ∠6   Interior Angles
∠1, ∠2, ∠7, ∠8   Exterior Angles

∠4, ∠5 }
∠3, ∠6 }   Alternate Interior Angles

∠1, ∠8 }
∠2, ∠7 }   Alternate Exterior Angles

∠1, ∠5 }
∠2, ∠6 }
∠3, ∠7 }   Corresponding Angles
∠4, ∠8 }

GEOMETRY

7. Triangles:
Sum of the angles in any triangle are equal to 180 degrees.

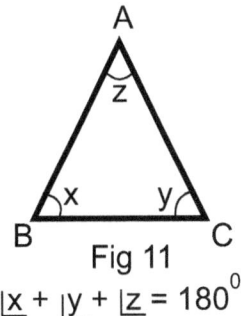

$\lfloor x + \lfloor y + \lfloor z = 180^0$

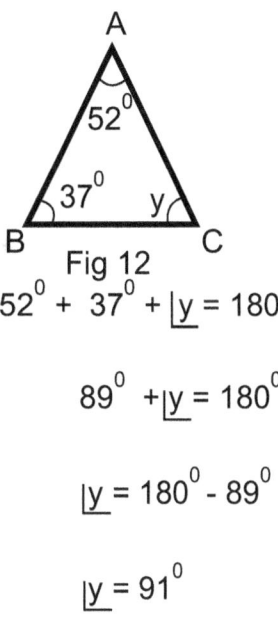

$52^0 + 37^0 + \lfloor y = 180^0$

$89^0 + \lfloor y = 180^0$

$\lfloor y = 180^0 - 89^0$

$\lfloor y = 91^0$

8. Quadrilaterals: Sum of the angles in any quadrilateral are equal to 360 degrees.

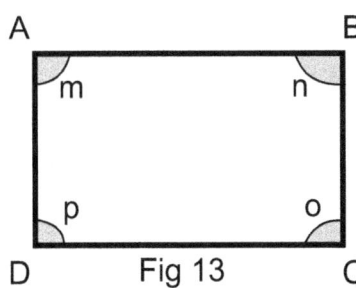

$\lfloor m + \lfloor n + \lfloor o + \lfloor p = 180^0$

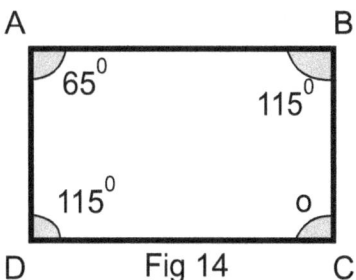

$\lfloor m + \lfloor n + \lfloor o + \lfloor p = 180^0$

$115^0 + 115^0 + 65^0 + \lfloor p = 180^0$

$295^0 + \lfloor p = 180^0$

$\lfloor p = 180^0 - 295^0$

$\lfloor p = 65^0$

9. Area of a triangle :

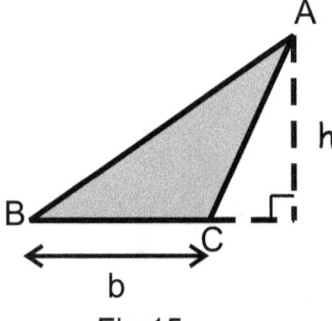
Fig 15

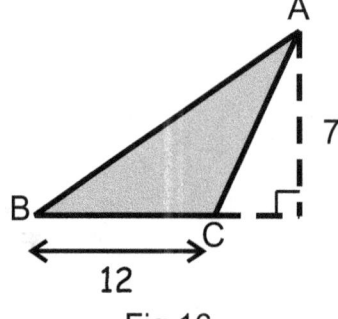
Fig 16

$A = \frac{1}{2} b h$

A = Area of the triangle
b = length of the base
h = height

$A = \frac{1}{2} b h$

$A = \frac{1}{2} \times (12)(7)$

$= (6)(7)$

$= 42$ sq. units

10. Perimeter of a triangle :

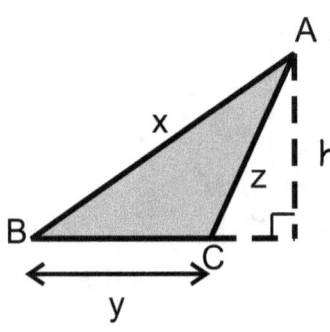
Fig 17

$P = x + y + z$
P = Perimeter
x, y, z are lengths of
the sides of the triangle

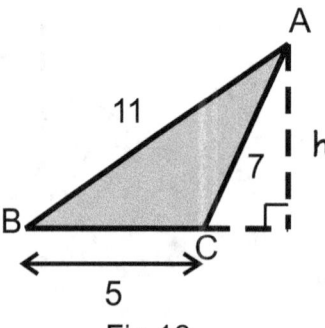
Fig 18

$P = x + y + z$
$P = 11 + 5 + 7$
$P = 23$ units

11. Perimeter and Area of a Square

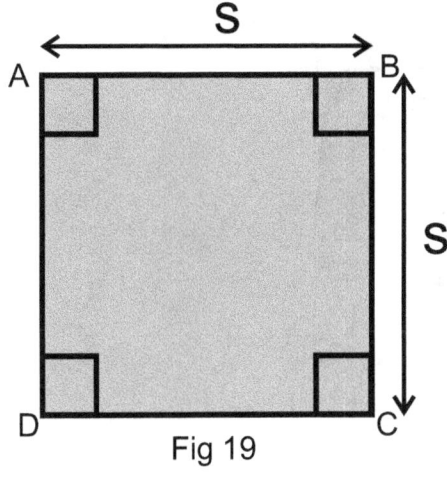
Fig 19

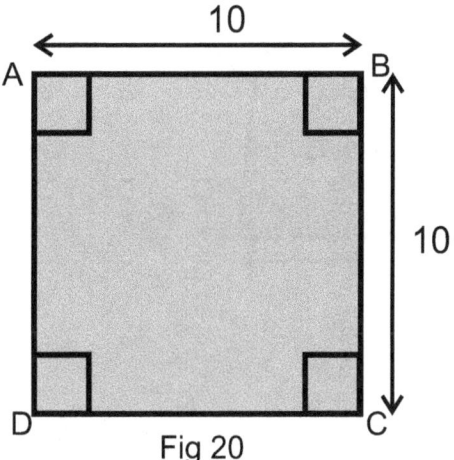
Fig 20

$P = 4s$
$A = s^2$

P = Perimeter
A = Area
s = Length of the side of the square

$P = 4s$
$P = 4 \times 10$
$P = 40$ units

$A = s^2$
$A = 10^2$
$A = 100$ sq.units

12. Perimeter and Area of a Rectangle

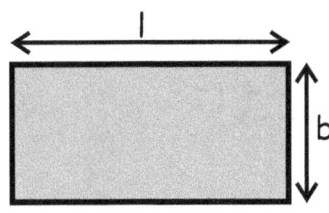
Fig 21

$A = l \times b$

$P = 2(l + b)$
A = Area
P = Perimeter
l = length of the rectangle
b = width of the rectangle

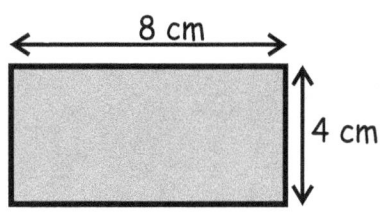
Fig 22

$A = 8 \times 4$
$A = 32$ sq.cm

$P = 2(8 + 4)$
$P = 2(12)$
$P = 24$ cm

13. Area of a Trapezium

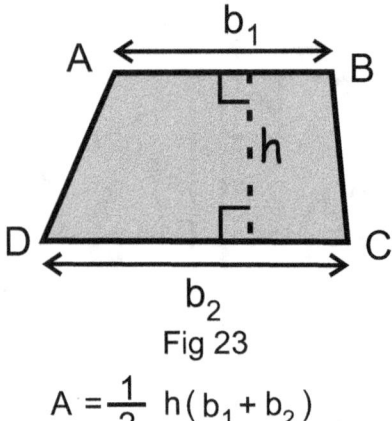

Fig 23

$A = \dfrac{1}{2} h(b_1 + b_2)$

A = Area
b_1, b_2 are the lengths of parallel sides
h = Distance between the parallel sides

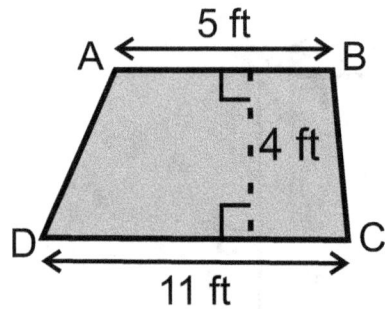

Fig 24

$A = \dfrac{1}{2} h(b_1 + b_2)$

$A = \dfrac{1}{2} 4(11 + 5)$

$A = 2(16)$

$A = 32$ sq.ft

14. Perimeter of a Trapezium

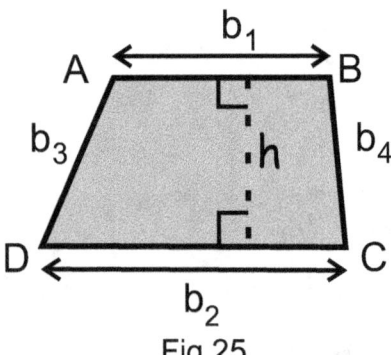

Fig 25

$P = b_1 + b_2 + b_3 + b_4$
P = Perimeter
b_1, b_2 are the lengths of parallel sides
b_3, b_4 are the lengths of non parallel sides

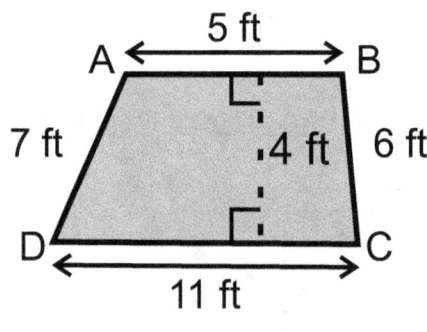

Fig 26

$P = b_1 + b_2 + b_3 + b_4$

$P = 5 + 11 + 7 + 6$

$P = 29$ ft

15. Area of a parellelogram

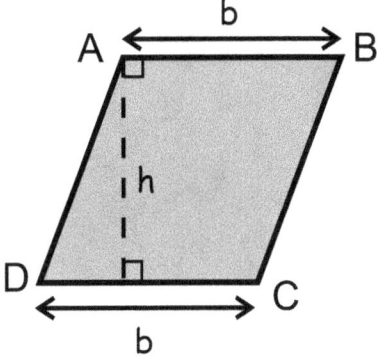

Fig 27

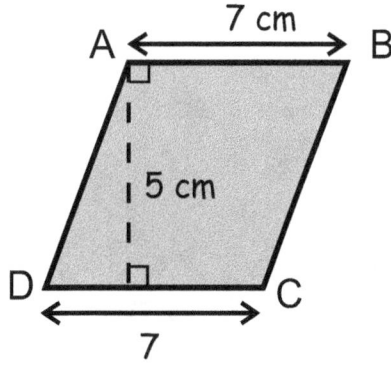

Fig 28

A = bh

A = Area

b = base

h = Height

A = bh

A = 5 X 7
A = 45 sq. cm

16. Perimeter of a parellelogram

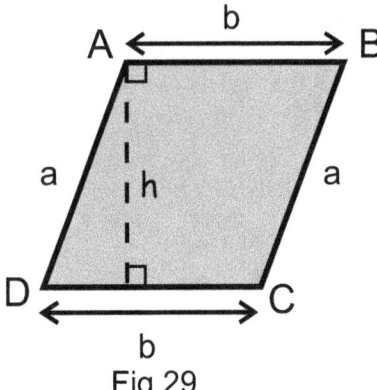

Fig 29

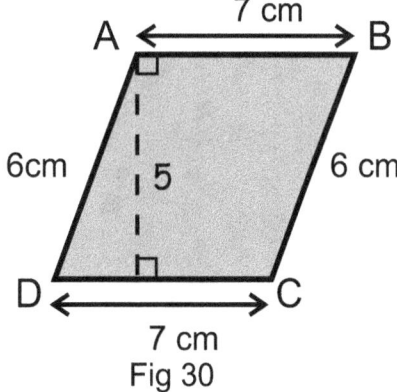

Fig 30

P = a + a + b + b
  = 2(a + b)

P = 6 + 6 + 7 + 7
  = 2(6 + 7)
  = 2(13)
  = 26 cm

17. Circumference and Area of a Circle

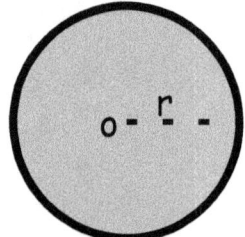

Fig 31

$C = \Pi r$

$A = \Pi r^2$

pi

$\Pi = 3.14$

$\Pi = \dfrac{22}{7}$

C = Circumference of the circle

A = Area of the circle
r = radius

Note : Diameter(d) = 2r

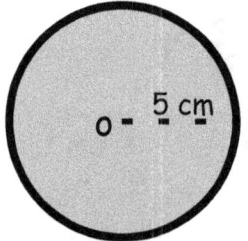

Fig 32

$C = 2\Pi r$

$C = 2 \times 3.14 \times 5$

$C = 10 \times 3.14$

$C = 31.4$ cm

$A = 3.14 \times 5 \times 5$
$A = 3.14 \times 25$
$A = 78.5$ sq.cm

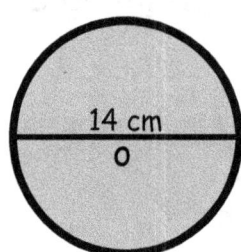

Fig 33

Diameter(d) = 14 cm

Radius = $\dfrac{14}{2}$

Radius = 7 cm

$C = 2 \times 3.14 \times 7$

$C = 14 \times 3.14$

$C = 43.86$ cm

$A = 3.14 \times 7 \times 7$
$A = 3.14 \times 49$
$A = 153.86$ sq.cm

18. Volume of a sphere

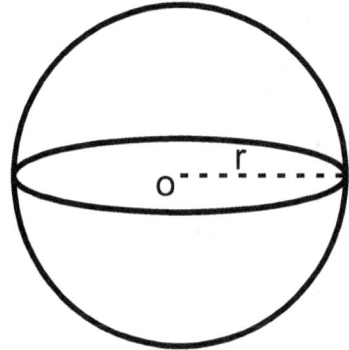

Fig 34

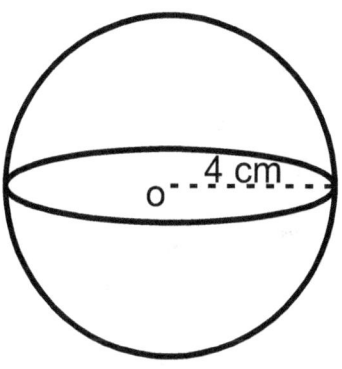

Fig 35

$V = \dfrac{4}{3} \Pi r^3$

V = Volume of the sphere
r = Radius of the sphere

Diameter = Twice the radius

pi

$\Pi = 3.14$

$\Pi = \dfrac{22}{7}$

$V = \dfrac{4}{3} \Pi r^3$

$V = \dfrac{4}{3} \times 3.14 \times 4^3$

$V = 268.08 \text{ cm}^3$

19. Surface area of a sphere

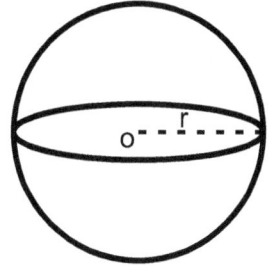

Fig 36

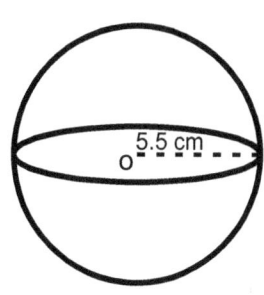

Fig 37

$S = 4\Pi r^2$

pi

$\Pi = 3.14$

$\Pi = \dfrac{22}{7}$

$S = 4\Pi r^2$

$S = 4 \times 3.14 \times (5.5)^2$

$S = 379.94 \text{ cm}^2$

20. Volume of a cone

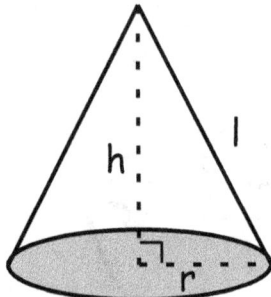

Fig 38

$V = \frac{1}{3} \Pi r^2 h$

V = Volume of the cone
r = Radius
h = Height
l = Slant height

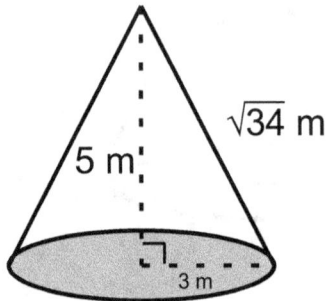

Fig 39

$V = \frac{1}{3} \Pi r^2 h$

$V = \frac{1}{3} \times 3.14 \times 3^2 \times 5$

$V = 47.1239 \ m^3$

21. Lateral surface area of a cone

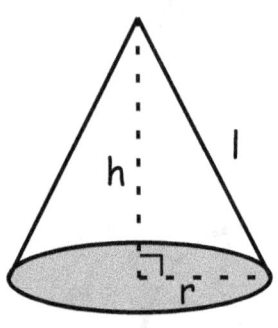

Fig 40

L.S.A $= \Pi r l$

$= \Pi r \times \sqrt{r^2 + h^2}$

$l = \sqrt{r^2 + h^2}$

L.S.A = Lateral surface area
l = Slant height of the cone
r = radius
h = height

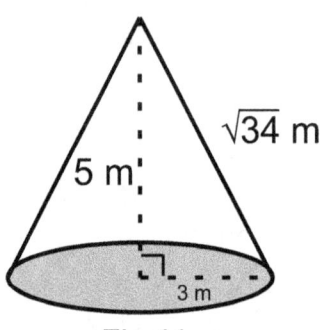

Fig 41

L.S.A $= \Pi r l$

L.S.A $= 3.14 \times 3 \times \sqrt{34}$

$= 54.9554 \ m^2$

GEOMETRY

Notes

22. Base surface area of a cone

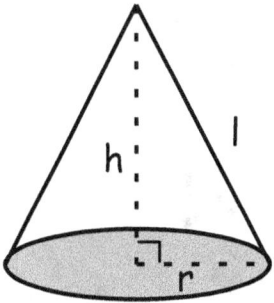

Fig 42

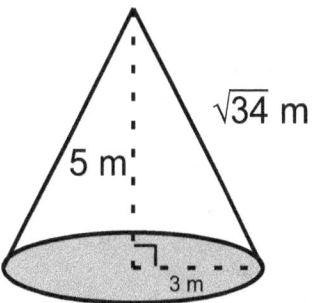

Fig 43

B.S.A $= \Pi r^2$

B.S.A = Base surface area
r = radius
h = height

B.S.A $= \Pi r^2$

B.S.A $= 3.14 \times 3^2$

B.S.A $= 28.2743 \text{ m}^2$

23. Total surface area of a cone

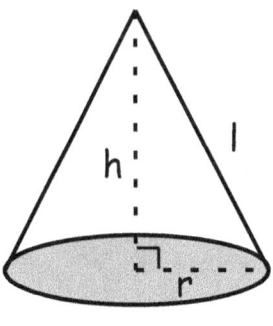

Fig 44

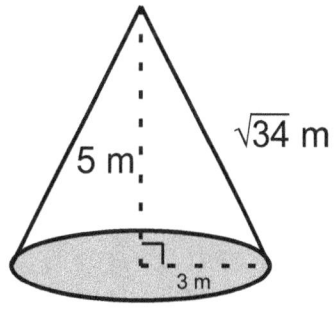

Fig 45

T.S.A = L.S.A + B.S.A

T.S.A $= \Pi rl + \Pi r^2$

T.S.A $= \Pi r(l + r)$

T.S.A $= \Pi r(r + \sqrt{r^2 + h^2})$

T.S.A = Total surface area
r = radius
h = height

T.S.A $= \Pi r(l + r)$

T.S.A $= 3.14 \times 3 (\sqrt{34} + 3)$

T.S.A $= 83.2298 \text{ m}^2$

24. Volume of a Cylinder

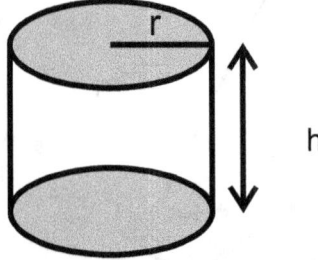

Fig 46

$V = \Pi r^2 h$

V = Volume
r = radius
h = height

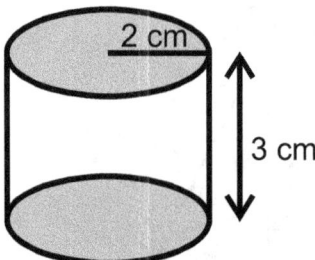

Fig 47

$V = \Pi r^2 h$

$V = 3.14 \times 2^2 \times 3$

$V = 37.6991 \text{ cm}^3$

25. Lateral Surface area of a Cylinder

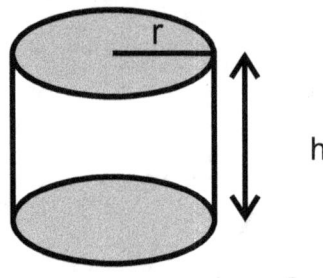

Fig 48

L.S.A $= 2 \Pi r h$

L.S.A = Lateral surface area
r = radius
h = height

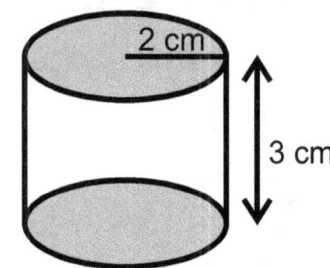

Fig 49

L.S.A $= 2 \Pi r h$

L.S.A $= 2 \times 3.14 \times 2 \times 3$

L.S.A $= 37.6991 \text{ cm}^2$

GEOMETRY

26. Top and bottom Surface area of a Cylinder

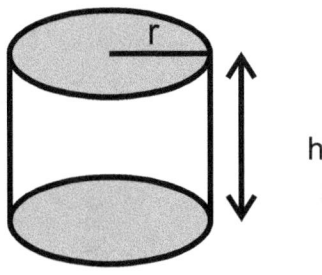

Fig 50

B.S.A = Πr^2

B.S.A = Bottom surface area
r = radius
h = height

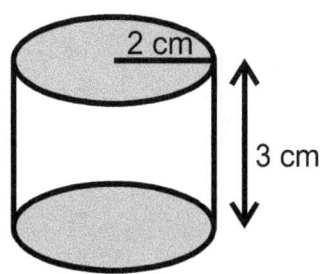

Fig 51

B.S.A = Πr^2

B.S.A = 3.14×2^2

B.S.A = 12.5664 cm^2

Top and bottom surface area of the cylinder are equal. While calculating total surface area of the cylinder, remember to add the bottom surface are of the cylinder twice

---

T.S.A = Total surface area
T.S.A = L.S.A + B.S.A + B.S.A

T.S.A = $2\Pi rh + \Pi r^2 + \Pi r^2$

T.S.A = $2\Pi rh + 2\Pi r^2$

T.S.A = $2\Pi r (h + r)$

T.S.A = $2\Pi r (h + r)$

T.S.A = $2 \times 3.14 \times 2 (3 + 2)$

T.S.A = 62.8318 cm^2

GEOMETRY

27. Volume of a Cuboid

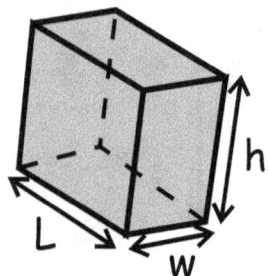

Fig 52

$V = lwh$
V = Volume
l = Length
w = Width or breadth
h = Height

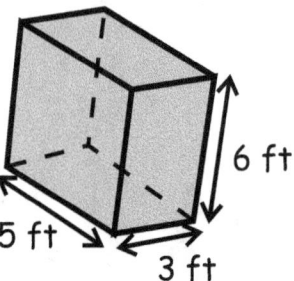

Fig 53

$V = lwh$

$V = 5 \times 3 \times 6$

$V = 90 \text{ ft}^3$

28. Surface area of a Cuboid

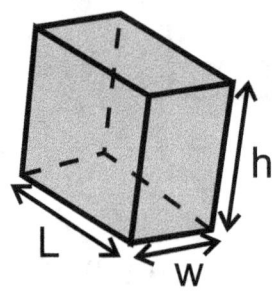

Fig 54

$T.S.A = 2(lw + lh + wh)$
T.S.A = Total surface area
l = Length
w = Width or breadth
h = Height

$L.S.A = 2(lh + wh)$
L.S.A = Lateral surface area
l = Length
w = Width or breadth
h = Height

Fig 55

$T.S.A = 2(lw + lh + wh)$

$T.S.A = 2(5 \times 3 + 5 \times 6 + 3 \times 6)$

$T.S.A = 126 \text{ ft}^2$

$L.S.A = 2(lh + wh)$

$L.S.A = 2(5 \times 6 + 3 \times 6)$

$L.S.A = 96 \text{ ft}^2$

29. Volume of a Cube

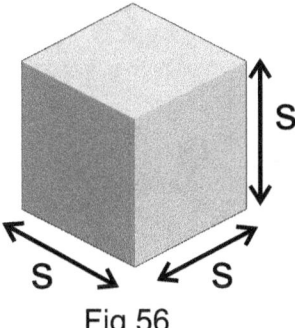

Fig 56

$V = s^3$
V = Volume
s = Side length of the cube

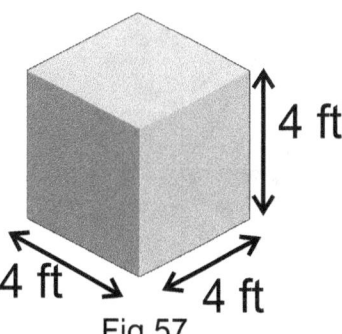

Fig 57

$V = s^3$
$V = 4^3$
$V = 64 \text{ ft}^3$

30. Surface area of a Cube

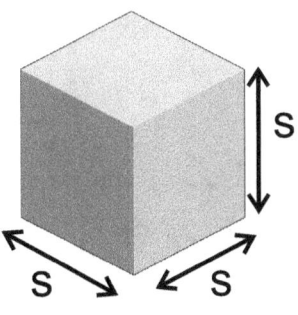

Fig 58

$T.S.A = 6s^2$
T.S.A = Total surface area
s = Side length of the cube

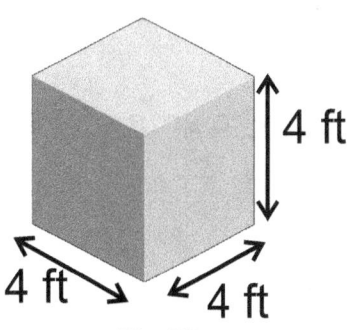

Fig 59

$T.S.A = 6s^2$
$T.S.A = 6 \times 4^2$
$T.S.A = 96 \text{ ft}^2$

31. Volume of a square Pyramid

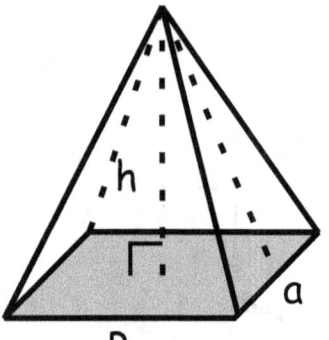
Fig 60

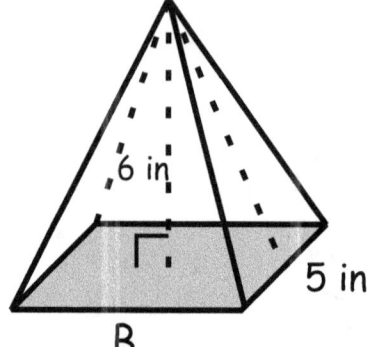
Fig 61

$V = \frac{1}{3} B h$

V = Volume
B = Base area
h = Height

$V = \frac{1}{3} a^2 h$

V = Volume
a = side length
h = Height

$V = \frac{1}{3} a^2 h$

$V = \frac{1}{3} \times 5^2 \times 6$

$V = \frac{1}{3} \times 150$

$V = 50 \text{ in}^3$

32. Lateral surface area of a square Pyramid

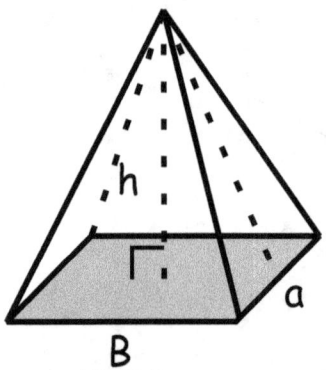
Fig 62

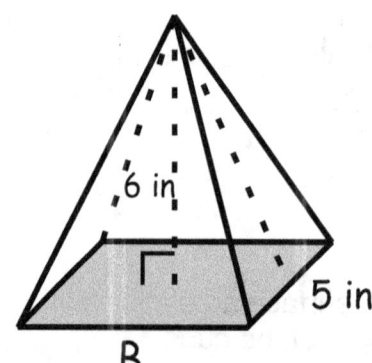
Fig 63

$S.A = a \times \sqrt{a^2 + 4h^2}$

$S.A = a \times \sqrt{a^2 + 4h^2}$

$S.A = 5 \times \sqrt{5^2 + 4 \times 6^2}$

$S.A = 65 \text{ in}^2$

32. Base surface area of a square Pyramid

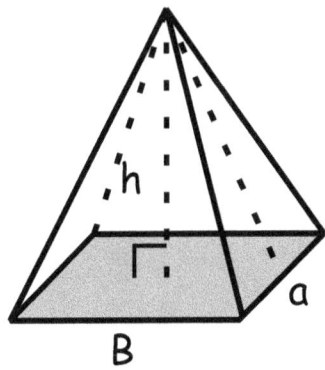

Fig 64

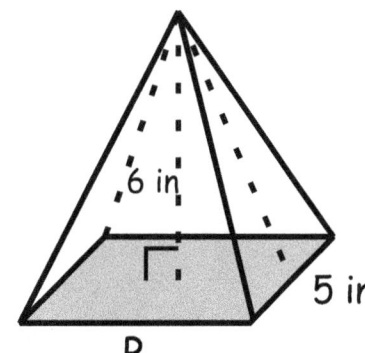

Fig 65

$B.S.A = a^2$

$T.S.A = a^2 + a \times \sqrt{a^2 + 4h^2}$

B.S.A = Base surface area
T.S.A = Total surface area
T.S.A = B.S.A + L.S.A

$B.S.A = 5 \times 5$

$B.S.A = 25\ in^2$

$T.S.A = 5^2 + 5 \times \sqrt{5^2 + 4 \times 6^2}$

$T.S.A = 25 + 65$

$T.S.A = 90\ in^2$

32. Volume of a Triangular prism

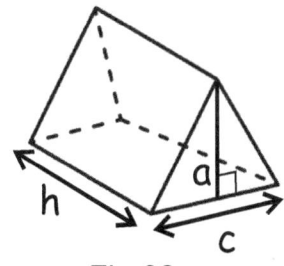

Fig 66

$V = Bh$

$V = \frac{1}{2} ach$

V = Volume
a = apothem
h = height
c = base length of the triangle

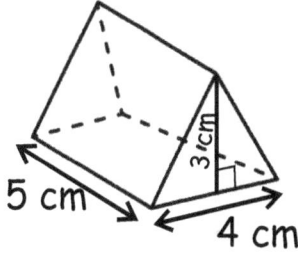

Fig 67

$V = Bh$

$V = \frac{1}{2} ach$

$V = \frac{1}{2}(3 \times 4 \times 5)$

$V = 30\ cm^3$

32. Volume of a triangular pyramid

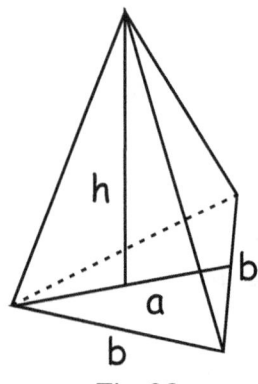

Fig 68

Volume of Triangular pyramid = $\frac{1}{6}$ abh

a = Apothem length of the pyramid
b = base length of the pyramid
h = Height of the pyramid

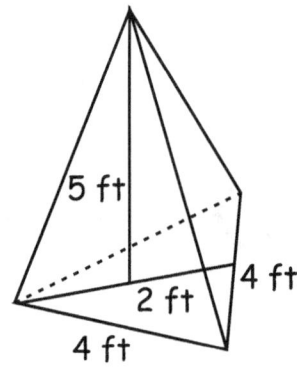

Fig 69

Volume of Triangular pyramid = $\frac{1}{6}$ abh

Volume of Triangular pyramid = $\frac{1}{6}$ (2X4X5)

Volume of Triangular pyramid = $\frac{1}{6}$ (40)

Volume of Triangular pyramid = 6.66 ft^3

33. Volume of a pentagonal pyramid

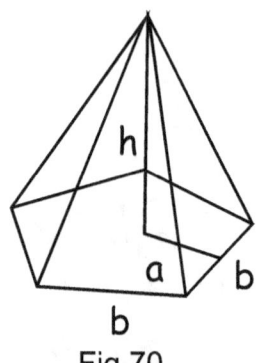

Fig 70

Volume of pentagonal pyramid = $\frac{5}{6}$ abh

a = Apothem length of the pyramid
b = base length of the pyramid
h = Height of the pyramid

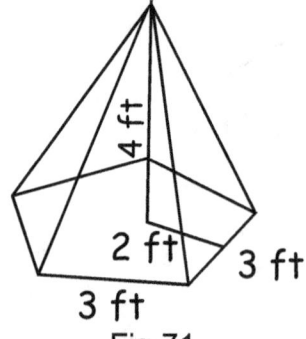

Fig 71

Volume of pentagonal pyramid = $\frac{5}{6}$ abh

Volume of pentagonal pyramid = $\frac{5}{6}$ (2X3X4)

Volume of pentagonal pyramid = $\frac{5}{6}$ (24)

Volume of pentagonal pyramid = 20 ft^3

34. Volume of a hexagonal pyramid

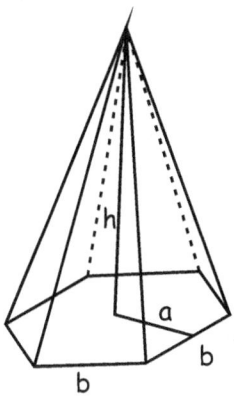

Fig 72

Volume of Hexagonal pyramid = abh

a = Apothem length of the pyramid
b = base length of the pyramid
h = Height of the pyramid

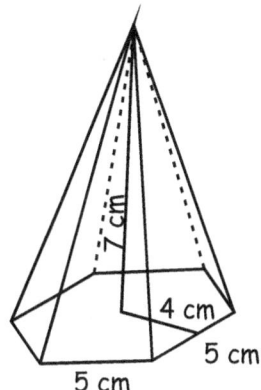

Fig 73

Volume of Hexagonal pyramid = 5X4X7
Volume of Hexagonal pyramid = 140 cm^3

FORMULAE

35. Equation of a Straight line

$y = mx + c$
Where m = slope
c = y - intercept

35. Two lines are parallel if and only if their slopes are equal.

36. Two lines are perpendicular if and only if product of their slopes equal to -1

37. Rotation of 90 (clockwise) Rotation of 270 (counter clockwise)

$(x, y) \longrightarrow (y, -x)$

38. Rotation of 90 (counter clockwise) Rotation of 270 (clockwise)

$(x, y) \longrightarrow (-y, x)$

39. Rotation of 180 (clockwise) & (counter clockwise)

$(x, y) \longrightarrow (-x, -y)$

40. Reflection over x-axis : $(x, y) \longrightarrow (x, -y)$

41.19. Reflection over y-axis : $(x, y) \longrightarrow (-x, y)$

42. Reflection over $x = y$: $(x, y) \longrightarrow (y, x)$

43. Reflection over $y = -x$: $(x, y) \longrightarrow (-y, -x)$

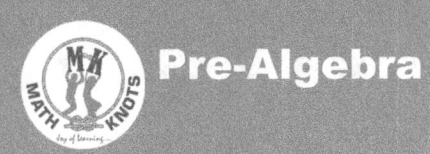

Pre-Algebra

FORMULAE

Abbreviations

milligram	mg	volume	V
gram	g	total Square Area	S.A
kilogram	kg	area of base	B
milliliter	mL	ounce	oz
liter	L	pound	lb
kiloliter	kL	quart	qt
millimeter	mm	gallon	gal.
centimeter	cm	inches	in.
meter	m	foot	ft
kilometer	km	yard	yd
square centimeter	cm^2	mile	mi.
cubic centimeter	cm^3	square inch	sq in.
		square foot	sq ft
		cubic inch	cu in.
		cubic foot	cu ft

year	yr
month	mon
hour	hr
minute	min
second	sec

FRACTIONS MULTIPLY #1

Pre-Algebra

FRACTIONS MULTIPLY #1

Find the product of the below fractions.

1. $2\frac{1}{6} \times -\frac{1}{5}$

(A) $\frac{13}{30}$ (B) $7\frac{17}{80}$ (C) $-\frac{13}{30}$ (D) $1\frac{29}{30}$

2. $-2\frac{1}{9} \times -2\frac{1}{4}$

(A) $-4\frac{3}{4}$ (B) 2 (C) -2 (D) $4\frac{3}{4}$

3. $5\frac{2}{5} \times -2\frac{1}{3}$

(A) $3\frac{1}{15}$ (B) $-18\frac{7}{20}$ (C) $-12\frac{3}{5}$ (D) $-16\frac{1}{2}$

4. $0 \times -\frac{7}{8}$

(A) $2\frac{5}{6}$ (B) $-\frac{3}{10}$ (C) 0 (D) $\frac{3}{10}$

Pre-Algebra Vol 2

FRACTIONS MULTIPLY #1

Find the product of the below fractions.

5. $-\dfrac{1}{2} \times \dfrac{2}{5}$

(A) $-1\dfrac{8}{15}$ (B) $-\dfrac{1}{5}$ (C) $-5\dfrac{18}{65}$ (D) $8\dfrac{4}{5}$

6. $-1\dfrac{8}{9} \times \dfrac{1}{5}$

(A) $-1\dfrac{31}{45}$ (B) $-3\dfrac{137}{210}$ (C) $3\dfrac{137}{210}$ (D) $-\dfrac{17}{45}$

7. $\dfrac{7}{4} \times -\dfrac{21}{13}$

(A) $-2\dfrac{43}{52}$ (B) $-\dfrac{439}{468}$ (C) $-3\dfrac{421}{572}$ (D) $1\dfrac{9}{52}$

8. $2 \times -\dfrac{15}{13}$

(A) $-2\dfrac{4}{13}$ (B) $-7\dfrac{97}{104}$ (C) $2\dfrac{4}{13}$ (D) $\dfrac{11}{13}$

FRACTIONS MULTIPLY #1

Find the product of the below fractions.

9. $-2\dfrac{3}{4} \times \dfrac{12}{7}$

(A) $1\dfrac{1}{15}$ (B) $-4\dfrac{5}{7}$ (C) $3\dfrac{9}{56}$ (D) $-1\dfrac{1}{15}$

10. $-3\dfrac{2}{5} \times 7\dfrac{4}{13}$

(A) $3\dfrac{59}{65}$ (B) $-23\dfrac{5}{52}$ (C) $-24\dfrac{11}{13}$ (D) $24\dfrac{11}{13}$

11. $-\dfrac{13}{8} \times -\dfrac{13}{9}$

(A) $11\dfrac{13}{72}$ (B) $\dfrac{173}{360}$ (C) $2\dfrac{25}{72}$ (D) $-2\dfrac{25}{72}$

12. $-\dfrac{7}{11} \times \dfrac{5}{8}$

(A) $-\dfrac{35}{88}$ (B) $\dfrac{7}{20}$ (C) $-\dfrac{7}{20}$ (D) $\dfrac{35}{88}$

Find the product of the below fractions.

13. $\boxed{-3\dfrac{1}{7} \times \dfrac{1}{3}}$

(A) $1\dfrac{1}{21}$ (B) $-9\dfrac{47}{105}$ (C) $-\dfrac{25}{84}$ (D) $-1\dfrac{1}{21}$

14. $\boxed{-13 \times \dfrac{11}{15}}$

(A) $-12\dfrac{2}{15}$ (B) $9\dfrac{8}{15}$ (C) $-9\dfrac{8}{15}$ (D) $-16\dfrac{8}{15}$

15. $\boxed{3\dfrac{1}{6} \times -3\dfrac{13}{14}}$

(A) $-12\dfrac{37}{84}$ (B) $-17\dfrac{155}{924}$ (C) $-13\dfrac{29}{42}$ (D) $12\dfrac{37}{84}$

FRACTIONS DIVIDE #2

Find each quotient of the below fractions.

1. $\boxed{1\dfrac{1}{10} \div \dfrac{4}{5}}$

   (A) $-\dfrac{8}{11}$ (B) $1\dfrac{3}{8}$ (C) $1\dfrac{9}{10}$ (D) $\dfrac{2}{10}$

2. $\boxed{\dfrac{1}{3} \div \dfrac{-9}{7}}$

   (A) $-\dfrac{7}{27}$ (B) $-\dfrac{5}{6}$ (C) $-3\dfrac{6}{7}$ (D) $-\dfrac{3}{7}$

3. $\boxed{\dfrac{4}{5} \div 4\dfrac{5}{7}}$

   (A) $5\dfrac{18}{35}$ (B) $-\dfrac{1}{6}$ (C) $\dfrac{28}{165}$ (D) $1\dfrac{10}{13}$

4. $\boxed{1 \div 2\dfrac{8}{5}}$

   (A) $-2\dfrac{1}{10}$ (B) $2\dfrac{1}{10}$ (C) $2\dfrac{7}{12}$ (D) $\dfrac{5}{18}$

Pre-Algebra Vol 2

FRACTIONS DIVIDE #2

Find each quotient of the below fractions.

5. $\dfrac{-8}{5} \div \dfrac{3}{2}$

(A) $-1\dfrac{1}{15}$ (B) $2\dfrac{1}{12}$ (C) $-\dfrac{1}{10}$ (D) $-3\dfrac{1}{10}$

6. $\dfrac{1}{14} \div 2$

(A) $-2\dfrac{7}{9}$ (B) $-\dfrac{1}{7}$ (C) $\dfrac{1}{28}$ (D) 28

7. $\dfrac{1}{4} \div -2\dfrac{2}{15}$

(A) $6\dfrac{1}{12}$ (B) $\dfrac{8}{15}$ (C) $-\dfrac{15}{128}$ (D) $8\dfrac{8}{15}$

8. $-1\dfrac{3}{10} \div -2$

(A) $\dfrac{13}{20}$ (B) $-\dfrac{13}{20}$ (C) 2 (D) $-\dfrac{2}{5}$

FRACTIONS DIVIDE #2

Find each quotient of the below fractions.

9. $\boxed{\dfrac{2}{3} \div 4\dfrac{5}{12}}$

   (A) $\dfrac{8}{53}$ (B) $-\dfrac{8}{53}$ (C) $6\dfrac{12}{13}$ (D) $-3\dfrac{3}{4}$

10. $\boxed{5\dfrac{5}{14} \div 1\dfrac{7}{8}}$

    (A) $3\dfrac{27}{56}$ (B) $2\dfrac{6}{7}$ (C) $\dfrac{7}{20}$ (D) $10\dfrac{5}{112}$

11. $\boxed{\dfrac{-1}{11} \div \dfrac{5}{9}}$

    (A) $-\dfrac{9}{55}$ (B) $-\dfrac{64}{99}$ (C) $-6\dfrac{1}{9}$ (D) $\dfrac{9}{55}$

12. $\boxed{5\dfrac{1}{6} \div \dfrac{-1}{2}}$

    (A) $-2\dfrac{7}{12}$ (B) $10\dfrac{1}{3}$ (C) $6\dfrac{7}{9}$ (D) $-10\dfrac{1}{3}$

Find each quotient of the below fractions.

13. $6\dfrac{1}{2} \div -1$

(A) $-\dfrac{2}{13}$ (B) $-6\dfrac{1}{2}$ (C) $5\dfrac{1}{2}$ (D) $6\dfrac{1}{2}$

14. $\dfrac{8}{5} \div \dfrac{-3}{14}$

(A) $1\dfrac{1}{15}$ (B) $-7\dfrac{7}{15}$ (C) $1\dfrac{27}{70}$ (D) $-1\dfrac{8}{9}$

15. $2 \div -1\dfrac{4}{9}$

(A) $1\dfrac{5}{8}$ (B) $\dfrac{5}{9}$ (C) $-1\dfrac{5}{13}$ (D) $2\dfrac{8}{9}$

Pre-Algebra Vol 2

FRACTIONS SUBTRACT #3

Simplify the below fractions.

1. $\boxed{24\dfrac{21}{22} - \dfrac{2}{3}}$

   (A) $25\dfrac{697}{759}$ (B) $24\dfrac{19}{66}$ (C) $44\dfrac{281}{1320}$ (D) $22\dfrac{397}{462}$

2. $\boxed{\dfrac{43}{39} - \dfrac{19}{21}}$

   (A) $5\dfrac{1501}{4368}$ (B) $25\dfrac{453}{910}$ (C) $\dfrac{18}{91}$ (D) $15\dfrac{1741}{2821}$

3. $\boxed{22\dfrac{1}{3} - \dfrac{7}{24}}$

   (A) $22\dfrac{1}{24}$ (B) $41\dfrac{953}{984}$ (C) $17\dfrac{1}{216}$ (D) $22\dfrac{1}{6}$

4. $\boxed{24\dfrac{19}{33} - 10\dfrac{19}{22}}$

   (A) $13\dfrac{47}{66}$ (B) $35\dfrac{152}{561}$ (C) $14\dfrac{103}{330}$ (D) $37\dfrac{289}{1518}$

FRACTIONS SUBTRACT #3

Simplify the below fractions.

5. $\boxed{9\dfrac{5}{9} - 7\dfrac{2}{5}}$

(A) $5\dfrac{496}{585}$ (B) $26\dfrac{43}{45}$ (C) $2\dfrac{7}{45}$ (D) $\dfrac{659}{765}$

6. $\boxed{\dfrac{29}{15} - \dfrac{3}{7}}$

(A) $2\dfrac{223}{2730}$ (B) $1\dfrac{53}{105}$ (C) $2\dfrac{677}{1050}$ (D) $3\dfrac{1}{210}$

7. $\boxed{23\dfrac{3}{47} - \dfrac{59}{34}}$

(A) $20\dfrac{36455}{62322}$ (B) $20\dfrac{14975}{15181}$ (C) $21\dfrac{525}{1598}$ (D) $19\dfrac{333}{799}$

8. $\boxed{10\dfrac{1}{6} - \dfrac{8}{23}}$

(A) $34\dfrac{995}{1173}$ (B) $10\dfrac{1064}{1173}$ (C) $30\dfrac{2329}{2760}$ (D) $9\dfrac{113}{138}$

FRACTIONS SUBTRACT #3

Simplify the below fractions.

9. $\boxed{19\dfrac{9}{17} - \dfrac{30}{19}}$

   (A) $40\dfrac{1391}{4522}$ (B) $17\dfrac{179}{2907}$ (C) $17\dfrac{307}{323}$ (D) $19\dfrac{889}{1615}$

10. $\boxed{7\dfrac{9}{10} - 2\dfrac{5}{18}}$

    (A) $5\dfrac{28}{45}$ (B) $6\dfrac{296}{765}$ (C) $6\dfrac{749}{1710}$ (D) $7\dfrac{28}{45}$

11. $\boxed{\dfrac{7}{5} - \dfrac{2}{17}}$

    (A) $\dfrac{1147}{1955}$ (B) $5\dfrac{1}{136}$ (C) $19\dfrac{1319}{1360}$ (D) $1\dfrac{24}{85}$

12. $\boxed{18\dfrac{1}{21} - \dfrac{9}{5}}$

    (A) $16\dfrac{967}{1785}$ (B) $15\dfrac{1543}{2415}$ (C) $16\dfrac{157}{210}$ (D) $16\dfrac{26}{105}$

Simplify the below fractions.

13. $15\dfrac{32}{43} - 12\dfrac{16}{33}$

(A) $3\dfrac{170}{1419}$ (B) $3\dfrac{368}{1419}$ (C) $4\dfrac{1735}{8514}$ (D) $25\dfrac{13193}{19866}$

14. $4\dfrac{1}{4} - \dfrac{31}{17}$

(A) $2\dfrac{1123}{1836}$ (B) $22\dfrac{29}{68}$ (C) $2\dfrac{29}{68}$ (D) $2\dfrac{65}{1428}$

15. $\dfrac{11}{6} - \dfrac{11}{10}$

(A) $1\dfrac{131}{165}$ (B) $19\dfrac{1}{30}$ (C) $\dfrac{11}{15}$ (D) $2\dfrac{22}{255}$

Evaluate the following.

1. $(-21.9) + 21.467 + (-8.9)$

   (A) -72.033   (B) 43.267   (C) -41.733   (D) -9.333

2. $55.1 + (-74.4) - (-90.3)$

   (A) 161.4   (B) 112.7   (C) 142.1   (D) 71

3. $58.73 + (-26.2) - (-44.66)$

   (A) 77.19   (B) 111.89   (C) 85.49   (D) 111.99

4. $79.4 + 65.9 - (-75.5)$

   (A) 273.29   (B) 247.6   (C) 182.3   (D) 220.8

5. $(75.5) - 5.6 + (-74.5)$

   (A) -4.6   (B) 81.5   (C) -38.8   (D) 49.1

Pre-Algebra Vol 2

DECIMALS ADD/SUBTRACT #4

Evaluate the following.

6. 44.3 − 48.7 − (−67)

   (A) 51.2   (B) 62.6   (C) 64.7   (D) −28

7. (−57.2) + (−99.7) − 8.1

   (A) −184.7   (B) −165   (C) −145.2   (D) −195

8. (−60.99) − 76.4 − 33

   (A) −150.25   (B) −170.39   (C) −135.39   (D) −123.49

9. (−41.7) − 77.6 + (−88.4)

   (A) −241   (B) −207.7   (C) −271.8   (D) −173.3

10. (−44.8) − 36.4 − (−84.8)

   (A) 49.4   (B) 3.6   (C) 9.1   (D) 42.2

DECIMALS ADD/SUBTRACT #4

Evaluate the following.

11. (−32.7) + (−85.2) − 73.7

   (A) −131   (B) −191.6   (C) −210.6   (D) −283.6

12. 75.5 − (−29.63) + (−62.3)

   (A) 20.54   (B) −2.17   (C) −2.57   (D) 42.83

13. (−15.5) + 73.9 + (−99)

   (A) −40.6   (B) −83   (C) −83.7   (D) −138

14. (−96.5) + (−3.3) − (−68.538)

   (A) 2.548   (B) −16.932   (C) −116.962   (D) −31.262

15. (−2) − 84.45 − 38.11

   (A) −223.56   (B) −124.56   (C) −173.66   (D) −36.86

Pre-Algebra Vol 2

DECIMALS MULTIPLY #5

Find the product of the below.

1. (-2.1)(-5.5)

   (A) 20.55   (B) -7.6   (C) 9.43   (D) 11.55

2. (8.4)(-3.4)

   (A) 28.56   (B) -32.06   (C) 5   (D) -28.56

3. (-4.4)(4.4)

   (A) -19.36   (B) -11.46   (C) 0   (D) -9.86

4. (-7.1)(-0.6)

   (A) -43.076   (B) 6.06   (C) -4.14   (D) 4.26

5. (5.2)(-1.7)

   (A) 8.84   (B) 3.5   (C) -16.349   (D) -8.84

Pre-Algebra Vol 2

DECIMALS MULTIPLY #5

Find the product of the below.

6. (-1.7)(-8.995)

   (A) -10.695   (B) 8.9915   (C) 15.2915   (D) -15.2915

7. (7.495)(-5.2)

   (A) -35.874   (B) 38.974   (C) -38.974   (D) 2.295

8. (-9.1)(4)

   (A) -34.3   (B) 36.4   (C) 34.86   (D) -36.4

9. (8.7)(-8.8)

   (A) -75.76   (B) 76.56   (C) -81.96   (D) -76.56

10. (-1.6)(6.2)

    (A) -8.42   (B) 9.92   (C) -18.321   (D) -9.92

DECIMALS MULTIPLY #5

Find the product of the below.

11. (-3)(-7.202)

(A) -10.202   (B) 15.306   (C) -21.606   (D) 21.606

12. (-4.3)(-8.6)

(A) 36.98   (B) 36.78   (C) -36.98   (D) 35.68

13. (-7.6)(9.3)

(A) 1.7   (B) -70.68   (C) -78.88   (D) 70.68

14. (-5.6)(-0.4)

(A) 10.54   (B) -54.52   (C) 2.24   (D) -2.24

15. (-8.3)(1.5)

(A) -6.8   (B) -4.35   (C) -12.45   (D) 12.45

Simplify the below.

1. $9 \div 3$

   (A) 3   (B) 7.573   (C) 12   (D) −9.1

2. $-3.3 \div -0.1$

   (A) −33   (B) 3.6   (C) 33   (D) −3.2

3. $-8.9 \div 2.5$

   (A) 5.9   (B) −11.4   (C) −3.56   (D) −8.81

4. $-6.6 \div 4$

   (A) 1.65   (B) −1.65   (C) 8.46   (D) −10.6

5. $5.7 \div 0.5$

   (A) 7.4   (B) 6.2   (C) 5.2   (D) 11.4

Simplify the below.

6. $\boxed{-5.7 \div 0.6}$

(A) -9.5 (B) 4.91 (C) -5.7 (D) 3.42

7. $\boxed{-5.4 \div 0.4}$

(A) -5 (B) -13.5 (C) 2.16 (D) 13.5

8. $\boxed{2.5 \div 2}$

(A) 1.25 (B) 1 (C) -0.8 (D) -8.4

9. $\boxed{-7.5 \div -2}$

(A) -15 (B) 3.75 (C) 5.7 (D) -9.5

10. $\boxed{-8.6 \div -2.5}$

(A) -3.7 (B) -5.5 (C) 3.44 (D) -9.9

Pre-Algebra Vol 2

DECIMALS DIVIDE #6

Simplify the below.

11. $-2.6 \div 5.2$

(A) 13.52 (B) -0.5 (C) 0.5 (D) 2

12. $-8 \div -0.1$

(A) -0.8 (B) -6.61 (C) -80 (D) 80

13. $5.4 \div -2$

(A) -4.7 (B) 10.8 (C) -2.7 (D) -9.88

14. $5.5 \div -2$

(A) 7.5 (B) 11 (C) 3.1 (D) -2.75

15. $-7.47 \div -0.3$

(A) 5.7 (B) -24.9 (C) -7.17 (D) 24.9

INTEGER MULTIPLICATION #7

Find the product of the below.

1. (−10)(4)(−4)

   (A) −108   (B) 160   (C) −160   (D) 108

2. (9)(−6)(−12)

   (A) −9   (B) 56   (C) 648   (D) −56

3. (−3)(6)(−5)

   (A) 90   (B) −90   (C) −288   (D) 288

4. (−8)(10)(7)

   (A) −560   (B) −90   (C) 560   (D) 90

5. (−7)(−1)(0)

   (A) −8   (B) 0   (C) −12   (D) −9

INTEGER MULTIPLICATION #7

Find the product of the below.

6. (13)(-3)(9)

   (A) -351   (B) 351   (C) 19   (D) -358

7. (6)(14)(-4)

   (A) -336   (B) -343   (C) 16   (D) -340

8. (-8)(-13)(-10)

   (A) -1034   (B) 1040   (C) -1040   (D) -31

9. (14)(11)(-12)

   (A) -1851   (B) -1855   (C) -1861   (D) -1848

10. (5)(10)(-7)

    (A) -357   (B) -340   (C) 350   (D) -350

INTEGER MULTIPLICATION #7

Find the product of the below.

11. (8)(10)(-4)

(A) -313   (B) -320   (C) 320   (D) 14

12. (-10)(-11)(6)

(A) -660   (B) 659   (C) 660   (D) -15

13. (12)(5)(-2)

(A) -130   (B) -120   (C) -131   (D) -113

14. (9)(-13)(-7)

(A) 819   (B) 280   (C) -280   (D) -819

15. (-11)(-5)(-10)

(A) 550   (B) -538   (C) -26   (D) -550

SCIENTIFIC NOTATION #8

Express the below in Scientific Notation.

1. $(4.6 \times 10^{-4})(5.3 \times 10^{0})$

   (A) 8.679×10^{-2} (B) 2.438×10^{-4}

   (C) 2.438×10^{-3} (D) 8.679×10^{-5}

2. $(7.2 \times 10^{-4})(7.9 \times 10^{-2})$

   (A) 5.688×10^{-4} (B) 9.114×10^{-2}

   (C) 9.114×10^{-3} (D) 5.688×10^{-5}

3. $(9.2 \times 10^{-5})(9.1 \times 10^{0})$

   (A) 83.72×10^{4} (B) 8.372×10^{-4}

   (C) 1.011×10^{-5} (D) 8.372×10^{4}

SCIENTIFIC NOTATION #8

Express the below in Scientific Notation.

4. $(2.73 \times 10^4)(3.8 \times 10^0)$

(A) 1.037×10^{-6} (B) 1.037×10^{-5}

(C) 0.1037×10^{-6} (D) 1.037×10^5

5. $(3.8 \times 10^{-6})(2.3 \times 10^4)$

(A) 8.74×10^2 (B) 8.74×10^3

(C) 8.74×10^{-2} (D) 1.652×10^{-10}

6. $(8.3 \times 10^{-5})(5 \times 10^1)$

(A) 4.15×10^{-4} (B) 4.15×10^{-3}

(C) 4.15×10^2 (D) 4.15×10^1

Express the below in Scientific Notation.

7. $(8 \times 10^0)(4.29 \times 10^{-2})$

(A) 3.432×10^{-2} (B) 3.432×10^{-1}

(C) 1.865×10^2 (D) 1.865×10^3

8. $(4.01 \times 10^1)(6.71 \times 10^3)$

(A) 5.976×10^{-3} (B) 2.691×10^6

(C) 59.76×10^{-3} (D) 2.691×10^5

9. $(7 \times 10^{-4})(4.5 \times 10^{-2})$

(A) 3.15×10^{-6} (B) 1.556×10^{-3}

(C) 3.15×10^{-5} (D) 1.556×10^{-2}

Express the below in Scientific Notation.

10. $(4.5 \times 10^6)(7.72 \times 10^{-6})$

(A) 3.474×10^0 (B) 3.474×10^{-1}
(C) 3.474×10^1 (D) 34.74×10^1

11. $(9.8 \times 10^1)(8.8 \times 10^{-1})$

(A) 1.114×10^2 (B) 86.24×10^1
(C) 862.4×10^1 (D) 8.624×10^1

12. $(8 \times 10^4)(6.74 \times 10^6)$

(A) 0.5392×10^{11} (B) 1.187×10^{-2}
(C) 5.392×10^{-11} (D) 5.392×10^{11}

Express the below in Scientific Notation.

13. $(2.87 \times 10^5)(4.4 \times 10^4)$

(A) 6.523×10^1 (B) 6.523×10^0

(C) 1.263×10^{10} (D) 0.6523×10^0

14. $(7.91 \times 10^5)(6.9 \times 10^{-6})$

(A) 5.458×10^{-1} (B) 54.58×10^0

(C) 54.58×10^1 (D) 5.458×10^0

15. $(5.3 \times 10^4)(6.09 \times 10^6)$

(A) 3.228×10^{10} (B) 3.228×10^{11}

(C) 8.703×10^{-2} (D) 8.703×10^{-3}

Pre-Algebra Vol 2

GCF #9

Find the Greatest Common Factor (GCF) of the below.

1. | 84, 72 |

   (A) 3   (B) 2   (C) 12   (D) 504

2. | 12, 88 |

   (A) 20   (B) 2   (C) 264   (D) 4

3. | 54, 72 |

   (A) 2   (B) 216   (C) 18   (D) 6

4. | 85, 34 |

   (A) 85   (B) 17   (C) 170   (D) 10

5. | 30, 70 |

   (A) 5   (B) 2   (C) 10   (D) 210

Find the Greatest Common Factor (GCF) of the below.

6. | 40, 72 |

(A) 8   (B) 16   (C) 360   (D) 2

7. | 48, 64 |

(A) 12   (B) 8   (C) 192   (D) 16

8. | 99, 54 |

(A) 594   (B) 9   (C) 4   (D) 3

9. | 40, 80 |

(A) 80   (B) 8   (C) 2   (D) 40

10. | 98, 21 |

(A) 7   (B) 14   (C) 21   (D) 294

GCF #9

Find the Greatest Common Factor (GCF) of the below.

11. | 52, 88 |

(A) 4   (B) 1144   (C) 2   (D) 3

12. | 56, 98 |

(A) 14   (B) 2   (C) 70   (D) 392

13. | 100, 50 |

(A) 100   (B) 37   (C) 25   (D) 50

14. | 50, 75 |

(A) 5   (B) 100   (C) 25   (D) 150

15. | 69, 92 |

(A) 23   (B) 276   (C) 4   (D) 92

Find the Greatest Common Factor (GCF) of the below.

1. 16, 48, 60v²

   (A) 240v²   (B) 2   (C) 4v   (D) 4

2. 68m, 68m², 34n

   (A) 17   (B) 34   (C) 68nm⁴   (D) 2

3. 50y, 40y, 65xy²

   (A) 5y   (B) 2y   (C) 3y   (D) 2600xy²

4. 80x³y, 40xy, 40y³x

   (A) 40xy   (B) 2xy   (C) 120xy   (D) 80y²x³

Pre-Algebra Vol 2

GCF MONOMIALS #10

Find the Greatest Common Factor (GCF) of the below.

5. $76, 38y^2, 76x^2$

   (A) 76   (B) $76x^4y^2$ (C) 31   (D) 38

6. $8x^2y, 76y^2x, 76x^2y$

   (A) $152x^2y^2$ (B) $4xy^2$ (C) $2xy$ (D) $4xy$

7. $78m, 52n, 54n^2$

   (A) $2m$ (B) 6   (C) $1404n^2m$ (D) 2

8. $60, 78y^2, 54y$

   (A) 2   (B) 6   (C) $6y$ (D) $7020y^2$

Find the Greatest Common Factor (GCF) of the below.

9. $75x^2y, 54y, 78x^2y$

   (A) 3y (B) 3x (C) 3yx (D) $17550x^2y$

10. $78n, 52n^3, 26mn$

    (A) $156mn^3$ (B) 13n (C) 2n (D) 26n

11. $50x^2, 50, 20$

    (A) 10 (B) 3 (C) 40 (D) $100x^2$

12. $42x^2y, 49x^2y, 49xy$

    (A) 3xy (B) 7xy (C) $294x^2y$ (D) 21xy

Find the Greatest Common Factor (GCF) of the below.

13. $\boxed{45y^2x^2,\ 27x^4,\ 45x^3}$

    (A) $9x^2$ (B) $135y^2x^4$ (C) $9x$ (D) $5x^2$

14. $\boxed{48x,\ 16y,\ 40}$

    (A) 8 (B) 5 (C) 2 (D) $240xy$

15. $\boxed{38u^3,\ 38u^3v,\ 57u^4}$

    (A) $76u^3$ (B) $19u^2$ (C) $114u^4v$ (D) $19u^3$

Pre-Algebra Vol 2

LCM NUMBERS #11

Find the Least Common Multiple (LCM) of the below.

1. | 42, 77 |

   (A) 3234   (B) 462   (C) 7   (D) 92

2. | 24, 90 |

   (A) 6   (B) 2160   (C) 15   (D) 360

3. | 63, 84 |

   (A) 167   (B) 252   (C) 21   (D) 5292

4. | 51, 78 |

   (A) 2   (B) 3978   (C) 1326   (D) 3

5. | 54, 63 |

   (A) 207   (B) 3402   (C) 378   (D) 9

Pre-Algebra Vol 2

LCM NUMBERS #11

Find the Least Common Multiple (LCM) of the below.

6. | 85, 68 |

(A) 17   (B) 340   (C) 20   (D) 5780

7. | 35, 100 |

(A) 5   (B) 3500   (C) 700   (D) 35

8. | 88, 86 |

(A) 2   (B) 3784   (C) 11   (D) 7568

9. | 72, 84 |

(A) 6048   (B) 12   (C) 422   (D) 504

10. | 67, 22 |

(A) 1474   (B) 223   (C) 1   (D) 134

Find the Least Common Multiple (LCM) of the below.

11. | 80, 48 |

(A) 3840   (B) 240   (C) 480   (D) 16

12. | 72, 96 |

(A) 6912   (B) 24   (C) 288   (D) 12

13. | 33, 39 |

(A) 3   (B) 1287   (C) 33   (D) 429

14. | 48, 64 |

(A) 192   (B) 3072   (C) 16   (D) 51

15. | 38, 95 |

(A) 38   (B) 190   (C) 3610   (D) 19

Pre-Algebra Vol 2
LCM MONOMIALS #12

Find the Least Common Multiple (LCM) of the below.

1. $88yx^2, 99yx^2, 55y^2x^2$

   (A) $3960y^2x^2$ (B) $479160y^4x^6$ (C) $90y^2x^2$ (D) $11yx^2$

2. $84x, 63x, 84$

   (A) $756x$ (B) $252x$ (C) $444528x^2$ (D) 21

3. $49x, 48, 67x$

   (A) 1 (B) $157584x^2$ (C) $157584x$ (D) $157584y$

4. $84y^2, 98y, 56x$

   (A) $1176xy^2$ (B) $168xy^2$ (C) $460992y^3x$ (D) 14

5. $72yx, 54x^2, 60x^3$

   (A) $6x$ (B) $30x^3y$ (C) $1080x^3y$ (D) $233280yx^6$

Find the Least Common Multiple (LCM) of the below.

6. 85u, 34u³, 34u²

   (A) 17u³   (B) 17u   (C) 170u³   (D) 98260u⁶

7. 36, 60a, 84

   (A) 181440a   (B) 12   (C) 2520a   (D) 1260a

8. 84ab, 42a², 42ab

   (A) 30ba²   (B) 42a   (C) 84ba²   (D) 148176a⁴b²

9. 66x, 59y, 63y²

   (A) 81774y²x²   (B) 81774y²x   (C) 1   (D) 245322xy³

10. 72a³, 48a, 96a

    (A) 288a³b   (B) 24a   (C) 288a³   (D) 331776a⁵

Find the Least Common Multiple (LCM) of the below.

11. $\boxed{85y^3, 68y, 85y^2}$

   (A) $340y^3$ (B) $17y$ (C) $491300y^6$ (D) $340y^3x$

12. $\boxed{46y^2, 69y^2, 92y^2}$

   (A) $276y^2$ (B) $292008y^6$ (C) $23y^2$ (D) $276y^2x$

13. $\boxed{93x, 60x^2, 69}$

   (A) 3 (B) $85560x^2$ (C) $385020x^3$ (D) $42780x^2$

14. $\boxed{24xy, 84, 84x^2}$

   (A) $169344x^{3y}$ (B) 12 (C) $168x^2y$ (D) $168x^3y$

15. $\boxed{81n^2m^2, 54m^3, 36m^2n}$

   (A) $2n^2m^3$ (B) $157464n^3m^8$ (C) $9m^2$ (D) $324n^2m^3$

ORDER OF OPERATIONS
#13

Evaluate.

1. $(19 \times 3 - 10 - (-9)) \div (-8)$

   (A) -14   (B) -6   (C) -13   (D) -7

2. $(18 - 4) \div (2(8 - 9))$

   (A) -2   (B) -5   (C) -7   (D) -14

3. $6 - ((-5) - 10 - 10) - 1$

   (A) 23   (B) 36   (C) 34   (D) 30

4. $((-3) - 5) \div ((-2)((-6) - (-8)))$

   (A) 1   (B) 3   (C) 2   (D) -8

5. $(4 - (6 + 5)) \times ((-7) \div (-7))$

   (A) -7   (B) -17   (C) -4   (D) -1

ORDER OF OPERATIONS #13

Evaluate.

6. $4 \div ((-3) + 1)^2 - 6$

　(A) -5　　(B) -14　　(C) 3　　(D) -11

7. $(-9) \div ((-9)((-2)^2 - 3))$

　(A) 4　　(B) 10　　(C) 7　　(D) 1

8. $(8 \times 2) \div (2 + (-4) - 6)$

　(A) -11　　(B) 2　　(C) 7　　(D) -2

9. $((-8) + 2) \times (-1) - ((-7) - 3)$

　(A) 17　　(B) 16　　(C) 26　　(D) 15

10. $18 \div (4 - (-2)) \times 25 \div (-5)$

　(A) -8　　(B) -18　　(C) -15　　(D) -11

ORDER OF OPERATIONS #13

Evaluate.

11. $((-7) - 5) \div ((-2) + 7 - 2)$

(A) –6   (B) –14   (C) –8   (D) –4

12. $(20 \div (4 \times (-1)^2)) \times 2$

(A) 0   (B) 10   (C) 6   (D) 13

13. $8 - 6 \div (3 + (-1) - 8)$

(A) –1   (B) 17   (C) 9   (D) 5

14. $7 + 5 \times (-4) - ((-10) - 2)$

(A) –1   (B) 2   (C) –5   (D) –6

15. $(-10) \times 6 - 7 \div (-1) \times 10$

(A) 5   (B) 13   (C) 10   (D) 18

Pre-Algebra Vol 2

VERBAL EXPRESSIONS #14

Express the below statement as an algebraic expression.

1. 4 less than n is equal to 22

   (A) $n - 4 = 22$ (B) $4n \leq 22$ (C) $4 - n = 22$ (D) $n^4 = 22$

2. A number cubed is less than or equal to 29

   (A) $n \cdot 3 \leq 29$ (B) $n + 3 \leq 29$ (C) $3^3 \leq 17$ (D) $n^3 \leq 29$

3. The quotient of a number and 2 is equal to 34

   (A) $\dfrac{n}{2} = 34$ (B) $\dfrac{2}{n}$ (C) $n - 2 = 34$ (D) $n + 2 = 34$

4. The n power of 15 is equal to 45

   (A) $15^n = 45$ (B) $15 + n = 45$ (C) $n - 15 = 45$ (D) $n^{15} = 45$

5. The 5th power of v is 40

   (A) $2 \cdot 5 = 40$ (B) $v^5 = 40$ (C) $5^v \leq 40$ (D) $5^v < 40$

Express the below statement as an algebraic expression.

6. The difference of k and 14 is less than 15

(A) $k + 14 < 15$ (B) $14 - 5$ (C) $k - 14 < 15$ (D) $\dfrac{k}{14} < 15$

7. The 3rd power of n is 20

(A) $n^3 = 20$ (B) $n + 3 < 20$ (C) $n + 3 = 20$ (D) $3^n = 20$

8. Half of a number is greater than 43

(A) $\dfrac{2}{2} > 43$ (B) $\dfrac{n}{2} > 43$ (C) $\dfrac{2}{2} \leq 43$ (D) $\dfrac{2}{2} \geq 43$

9. q squared is equal to 42

(A) $2^2 = 42$ (B) $\dfrac{q}{2} = 42$ (C) $q + 2 = 42$ (D) $q^2 = 42$

10. 7 to the n is equal to 22

(A) $n^7 \geq 22$ (B) $7^n = 22$ (C) $n - 7 = 22$ (D) $n^7 \leq 22$

VERBAL EXPRESSIONS #14

Express the below statement as an algebraic expression.

11. | 11 more than r is equal to 40 |

(A) $11^r = 6$ (B) $\dfrac{11}{r} = 40$ (C) $r^3 = 40$ (D) $r + 11 = 40$

12. | 4 to the d is less than 16 |

(A) $4^d < 16$ (B) $d - 4 = 16$ (C) $d + t < 16$ (D) $4 - d < 16$

13. | A number decreased by 28 is greater than 18 |

(A) $n - 28 > 18$ (B) $n^2 > 18$ (C) $n^3 = 18$ (D) $\dfrac{n}{28} > 18$

14. | The product of q and 11 is less than 19 |

(A) $q^{11} < 19$ (B) $\dfrac{q}{11} < 19$ (C) $q - 11 < 19$ (D) $q \cdot 11 < 19$

15. | 22 less than x is 29 |

(A) $x - 22 = 29$ (B) $22^x < 29$ (C) $x^3 > 29$ (D) $22^2 = 29$

VERBAL EXPRESSION EQUATION #15

Write the verbal expression of the inequation.

1. $\dfrac{n}{6} = 32$

   (A) The quotient of a number and 6 is 32.
   (B) A number minus 6 is 32.
   (C) The quotient of 6 and a number is 32.
   (D) The difference of 6 and a number is 32.

2. $n - 10 < 36$

   (A) Twice a number is less than 36.
   (B) 10 minus a number is less than 36.
   (C) A number times 10 is less than 36.
   (D) A number minus 10 is less than 36.

3. $n + 7 \geq 28$

   (A) A number to the 7th is greater than or equal to 28.
   (B) A number less than 7 is greater than or equal to 28.
   (C) The difference of 7 and a number is greater than or equal to 28.
   (D) The sum of a number and 7 is greater than or equal to 28.

Write the verbal expression of the inequation.

4. $n - 17 = 11$

    (A) 17 cubed is 11.
    (B) 17 plus n is 11.
    (C) n less than 17 is 11.
    (D) 17 less than n is 11.

5. $n - 24 \leq 33$

    (A) n cubed is less than or equal to 33.
    (B) n less than 24 is less than or equal to 33.
    (C) The quotient of n and 24 is less than or equal to 33.
    (D) 24 less than n is less than or equal to 33.

6. $24 - 19$

    (A) The difference of 19 and 24.
    (B) 24 divided by 19.
    (C) The difference of 24 and 19.
    (D) 19 increased by 24.

Write the verbal expression of the inequation.

7. n^7

(A) The difference of n and 7.
(B) 7 more than n.
(C) The 7th power of n.
(D) The quotient of n and 7.

8. $3n$

(A) Half of 3.
(B) A number increased by 3 is less than 3.
(C) A number decreased by 3.
(D) Three times a number

9. $\frac{n}{8} \geq 21$

(A) 8 less than a number is greater than or equal to 21.
(B) A number less than 8 is greater than or equal to 21.
(C) A number divided by 8 is greater than or equal to 21.
(D) A number cubed is greater than or equal to 21.

Write the verbal expression of the inequation.

10. $n^9 \geq 32$

    (A) 9 decreased by n is greater than or equal to 32.
    (B) The 9th power of n is greater than or equal to 32.
    (C) The n power of 9 is greater than or equal to 32.
    (D) n plus 9 is greater than or equal to 32.

11. $n + 5 = 16$

    (A) The quotient of 5 and a number is 16.
    (B) The sum of a number and 5 is 16.
    (C) A number minus 5 is 16.
    (D) Twice a number is 16.

12. $2n \geq 5$

    (A) 2 squared is greater than or equal to 5.
    (B) The sum of a number and 2 is greater than or equal to 5.
    (C) A number less than 2 is greater than or equal to 5.
    (D) Twice a number is greater than or equal to 5.

VERBAL EXPRESSION EQUATION #15

Write the verbal expression of the inequation.

13. $\boxed{2n}$

    (A) A number squared.
    (B) Half of a number.
    (C) 2 to the n.
    (D) Twice a number.

14. $\boxed{n \cdot 7}$

    (A) 7 divided by a number.
    (B) A number divided by 7.
    (C) A number to the 7th.
    (D) The product of a number and 7.

15. $\boxed{v^4}$

    (A) The 4th power of v.
    (B) The v power of 4.
    (C) v less than 4.
    (D) The product of v and 4.

Pre-Algebra Vol 2

MONOMIALS #16

Simplify the below expression.

1. $-3n - 5(n - 7)$

   (A) $-7n + 40$ (B) $6 - n$ (C) $-8n + 43$ (D) $-8n + 35$

2. $-8k - 5(k + 6)$

   (A) $-2 - 8k$ (B) $-3 - 8k$ (C) $-13k - 30$ (D) $17k + 30$

3. $-3n - 3(1 - n)$

   (A) $-9n + 36$ (B) $-9n + 30$ (C) -3 (D) $-9n + 32$

4. $6 - 8(m - 2)$

   (A) $19 - 8m$ (B) $-6m + 37$ (C) $22 - 8m$ (D) $-6m + 39$

5. $-4 - 4(8n - 4)$

   (A) $12 - 32n$ (B) $17n - 32$ (C) $19n - 32$ (D) $-16 + 16n$

MONOMIALS #16

Simplify the below expression.

6. $6v - (7v + 5)$

   (A) $-2 - 25v$ (B) $-5v - 12$ (C) $-5v - 16$ (D) $-v - 5$

7. $-(-p - 1) - 2p$

   (A) $-2p + 2$ (B) $-p + 1$ (C) $-p - 5$ (D) $-2p + 5$

8. $-5 - 8(-1 + 6m)$

   (A) $3 - 48m$ (B) $2m - 42$ (C) $9m - 21$ (D) $13 - 5m$

9. $2x - 3(6x + 7)$

   (A) $-5x - 25$ (B) $-5x - 20$ (C) $-16x - 21$ (D) $-15x - 9$

10. $-8(5x + 8) + 6x$

    (A) $-10x + 18$ (B) $-36x - 64$ (C) $-34x - 64$ (D) $-39x - 64$

MONOMIALS #16

Simplify the below expression.

11. $-7(8n - 5) - 8$

   (A) $-3n + 32$ (B) $-56n + 20$ (C) $20 + 27n$ (D) $-56n + 27$

12. $-2(4a - 3) - 8a$

   (A) $-11a + 13$ (B) $-20 + 22a$ (C) $-16a + 6$ (D) $-16a + 13$

13. $-8 - 5(1 + 5x)$

   (A) $-5 - 25x$ (B) $-13 - 25x$ (C) $-10x - 18$ (D) $-21x - 6$

14. $-5(6x + 5) + x$

   (A) $-5 - 8x$ (B) $-12 - 8x$ (C) $-29x - 25$ (D) $-28 - 30x$

15. $-6(5 + 8k) - 8$

   (A) $38k - 8$ (B) $-31 - 48k$ (C) $38k - 11$ (D) $-38 - 48k$

INEQUALITIES #17

Solve the below inequality.

1. $4(-7n - 1) + n \leq 131$

   (A) {All real numbers.} (B) $n \leq -5$
   (C) $n \leq -9$ (D) $n \geq -5$

2. $-5(3 + 7v) - 4v > -171$

   (A) {All real numbers.} (B) $v < -9$
   (C) $v < 4$ (D) $v > -9$

3. $-4(7b + 1) > 220$

   (A) $b > -21$ (B) $b < -21$
   (C) $b > -8$ (D) $b < -8$

4. $-215 \geq 2 + 7(1 - 4n)$

   (A) $n \geq 4$ (B) $n \geq 8$
   (C) $n \geq -14$ (D) $n \geq -12$

INEQUALITIES #17

Solve the below inequality.

5. $-90 > 5(5x + 7)$

(A) $x < -5$ (B) $x < -38$
(C) $x < -26$ (D) {All real numbers.}

6. $-3(5n + 3) < 111$

(A) $n < -13$ (B) $n < -8$
(C) $n > -8$ (D) $n < -27$

7. $-88 > 8(n - 6) + 2n$

(A) $n < -4$ (B) $n > -37$
(C) $n < -22$ (D) $n = -4$

8. $84 > -6(7m + 7)$

(A) $m > -7$ (B) $m < -19$
(C) $m > -19$ (D) $m > -3$

INEQUALITIES #17

Solve the below inequality.

9. $-8(3m + 5) \leq -232$

(A) $m \leq -6$ (B) $m \leq 8$
(C) $m \geq 8$ (D) $m \geq -6$

10. $5(1 - 6n) > -85$

(A) $n < 3$ (B) $n < -17$
(C) $n > -9$ (D) $n < -9$

11. $6(3 - 5n) < 228$

(A) $n > -40$ (B) $n > -21$
(C) $n > -7$ (D) $n > -16$

12. $112 \leq 4 + 3(5x - 4)$

(A) $x \leq -12$ (B) $x \geq 8$
(C) $x \geq -12$ (D) $x \leq -40$

INEQUALITIES #17

Solve the below inequality.

13. $88 \leq 8(x + 7)$

(A) $x \leq -28$ (B) $x \leq 4$
(C) $x \geq 4$ (D) $x \leq -5$

14. $3(6v - 6) \leq -108$

(A) $v \geq -36$ (B) {All real numbers.}
(C) $v \leq -5$ (D) $v \leq -36$

15. $-5(-3v - 7) \geq -140$

(A) $v \leq -9$ (B) $v \geq -7$
(C) $v \leq 5$ (D) $v \leq -7$

Evaluate the below expression.

1. $7|y| + |x|$; use $x = 10$, and $y = 7$

    (A) 53 (B) 59 (C) 58 (D) 65

2. $b - \dfrac{|b + a|}{3}$; use $a = 6$, and $b = 9$

    (A) -3 (B) 1 (C) 5 (D) 4

3. $3x + y + |y|$; use $x = -1$, and $y = -3$

    (A) -1 (B) -3 (C) -5 (D) -12

4. $-\dfrac{5}{5} \times |y - x|$; use $x = 4$, and $y = -8$

    (A) -2 (B) -12 (C) -10 (D) -22

5. $b - 10 - \left(\dfrac{8}{4} + a\right)$; use $a = 10$, and $b = -10$

    (A) -24 (B) -26 (C) -32 (D) -30

Evaluate the below expression.

6. $x(y + x^2) - y$; use $x = -1$, and $y = 8$

(A) –22 (B) –18 (C) –17 (D) –8

7. $p + pr(-5 - q)$; use $p = -1$, $q = 9$, and $r = -5$

(A) –80 (B) –71 (C) –79 (D) –63

8. $m + m - n^2 - n$; use $m = 4$, and $n = -5$

(A) –12 (B) –9 (C) –14 (D) –16

9. $(z^2)^2 + x + z$; use $x = -4$, and $z = -3$

(A) 75 (B) 70 (C) 84 (D) 74

10. $mn - (n + m - 4)$; use $m = -2$, and $n = 5$

(A) –11 (B) –9 (C) –13 (D) –15

EVALUATE EXPRESSIONS #18

Evaluate the below expression.

11. $b - (|a + b| - 4)$; use $a = -7$, and $b = -8$

(A) -19   (B) 39   (C) 27   (D) 19

12. $P(m + p + p + 5)$; use $m = 1$, and $p = -5$

(A) 29   (B) 20   (C) 11   (D) 13

13. $x^2 + x + y - x$; use $x = 7$, and $y = -1$

(A) 48   (B) 56   (C) 51   (D) 58

14. $\dfrac{p - 3 + 7 + n}{3}$; use $n = 8$, and $p = 3$

(A) 12   (B) 5   (C) -5   (D) -1

15. $\dfrac{xy}{4} + |x|$; use $x = 10$, and $y = -10$

(A) -15   (B) -10   (C) -11   (D) -8

Pre-Algebra Vol 2

EVALUATE EXPRESSIONS #19

Evaluate the below expression.

1. $-24 - n = 29$

   (A) -53 (B) $-1\frac{5}{24}$ (C) 5 (D) 53

2. $85 = \frac{n}{21}$

   (A) 1785 (B) 62 (C) $4\frac{1}{21}$ (D) 64

3. $-33 = a + (-16)$

   (A) $2\frac{1}{16}$ (B) -17 (C) -49 (D) 14

4. $a + (-63) = -60$

   (A) $\frac{20}{21}$ (B) -13 (C) -21 (D) 3

5. $84x = 4956$

   (A) 59 (B) 8 (C) -36 (D) 93

Pre-Algebra Vol 2

EVALUATE EXPRESSIONS #19

Evaluate the below expression.

6. $166 = 91 + k$

(A) 75   (B) 3   (C) 38   (D) 85

7. $-26m = -2470$

(A) −60   (B) 83   (C) 95   (D) −21

8. $-135 = 27r$

(A) 36   (B) 89   (C) −5   (D) 64

9. $49 = m - 5$

(A) −35   (B) 54   (C) $9\frac{4}{5}$ (D) 44

10. $6882 = 74m$

(A) −39   (B) −58   (C) −25   (D) 93

EVALUATE EXPRESSIONS #19

Evaluate the below expression.

11. $\dfrac{67}{34} = \dfrac{x}{34}$

(A) 67   (B) 32   (C) 40   (D) -54

12. $-20 = m + (-53)$

(A) -73   (B) 33   (C) $\dfrac{20}{53}$ (D) 74

13. $\dfrac{k}{12} = -96$

(A) -39   (B) -1152   (C) -84   (D) -8

14. $-3 = k - 78$

(A) $-\dfrac{1}{26}$ (B) -18   (C) -81   (D) 75

15. $-2145 = -65b$

(A) -37   (B) 33   (C) -11   (D) 75

SOLVE FOR X #20

Solve the below equation.

1. $7(-3n + 6) = 126$

   (A) −8 (B) −4
   (C) 12 (D) All real numbers

2. $8(8x - 7) = -120$

   (A) −2 (B) −15
   (C) −1 (D) 14

3. $-6(-2r - 3) = 90$

   (A) −12 (B) All real numbers
   (C) −6   (D) 6

4. $-174 = 8(3x - 3) + 6x$

   (A) −5 (B) −1
   (C) 6   (D) No solution

Pre-Algebra Vol 2

SOLVE FOR X #20

Solve the below equation.

5. $7(2v + 6) = 154$

(A) −10 (B) 2
(C) 10 (D) 8

6. $-496 = 8(8m - 6)$

(A) −7 (B) −8
(C) −14 (D) 6

7. $4(2x - 5) = -84$

(A) −8 (B) 15
(C) 9 (D) 5

8. $126 = 5(6 - 7k) + 3k$

(A) 7 (B) −3
(C) −5 (D) 1

Solve the below equation.

9. $144 = 2(8 + 8n)$

   (A) 16 (B) 8
   (C) 10 (D) 1

10. $150 = 5(4x + 2)$

    (A) 7 (B) 8
    (C) All real numbers (D) 6

11. $115 = 5(3n + 8)$

    (A) 5 (B) −9
    (C) 1 (D) 6

12. $264 = -8(7 - 5k)$

    (A) 15 (B) 3
    (C) 8 (D) No solution

Solve the below equation.

13. $8(1 - 7x) = 176$

   (A) 6 (B) −3
   (C) −4 (D) 12

14. $-246 = -8(4n + 8) + 6n$

   (A) 7 (B) 10
   (C) −15 (D) −4

15. $90 = 5(2x + 6)$

   (A) 9 (B) 6
   (C) 8 (D) 16

Evaluate Expressions #21

Evaluate the below expressions using the values given.

1. $8 - r + p$; use $p = -1$, and $r = -10$

   (A) 12   (B) 17   (C) 22   (D) 19

2. $j - 5h$; use $h = 6$, and $j = 6$

   (A) -24   (B) -34   (C) -20   (D) -32

3. $z - (x + y)$; use $x = 7$, $y = -5$ and $z = -6$

   (A) -16   (B) 0   (C) -13   (D) -8

4. $y + 7 - x$; use $x = -4$, and $y = 6$

   (A) 20   (B) 17   (C) 7   (D) 10

5. $p + 5m$; use $m = -6$, and $p = -2$

   (A) -32   (B) -37   (C) -28   (D) -30

Pre-Algebra Vol 2

Evaluate Expressions #21

Evaluate the below expressions using the values given.

6. $j + jh$; use $h = 5$, and $j = -2$

   (A) -9   (B) -12   (C) -10   (D) -13

7. $j + |h|$; use $h = -7$, and $j = 5$

   (A) 3   (B) 21   (C) 20   (D) 12

8. $y + y + x$; use $x = 2$, and $y = -3$

   (A) -4   (B) -14   (C) -10   (D) -9

9. $m^2 p$; use $m = 4$, and $p = 2$

   (A) 26   (B) 32   (C) 33   (D) 23

10. $y - (x + y)$; use $x = -7$, and $y = 7$

    (A) 9   (B) 5   (C) 7   (D) 10

Pre-Algebra Vol 2

Evaluate Expressions #21

Evaluate the below expressions using the values given.

11. $a(b + 7)$; use $a = -1$, and $b = 4$

   (A) −11 (B) −21 (C) −8 (D) −14

12. $z + \dfrac{x}{3}$; use $x = -3$, and $z = 2$

   (A) 5 (B) −8 (C) 1 (D) 8

13. $7(m - q)$; use $m = -7$, and $q = 1$

   (A) −58 (B) −54 (C) −48 (D) −56

14. $x - (y - y)$; use $x = 4$, and $y = -6$

   (A) 4 (B) 6 (C) −6 (D) 2

15. $b - (a - a)$; use $a = -10$, and $b = 1$

   (A) 1 (B) −6 (C) 5 (D) −8

ABSOLUTE VALUE #22

Solve the below equation.

1. $|5r - 10| = 15$

   (A) $\{5, -1\}$ (B) $\{1, 0\}$

   (C) $\left\{-5, \dfrac{31}{5}\right\}$ (D) $\{5\}$

2. $|-6k - 9| = 21$

   (A) $\{-5\}$ (B) $\{-5, 2\}$

   (C) $\left\{\dfrac{4}{5}\right\}$ (D) $\left\{\dfrac{4}{5}, -4\right\}$

3. $|2r + 3| = 11$

   (A) $\{-1, -9\}$ (B) $\{4, -7\}$

   (C) $\{5, -3\}$ (D) $\{4\}$

Solve the below equation.

4. $|1 - 7p| = 57$

(A) $\left\{-\dfrac{1}{2}, 1\right\}$ (B) $\{6, -9\}$

(C) $\{6\}$ (D) $\left\{-8, \dfrac{58}{7}\right\}$

5. $|6 + 7a| = 36$

(A) $\{2, 3\}$ (B) $\left\{\dfrac{30}{7}, -6\right\}$

(C) $\{4\}$ (D) $\left\{4, -\dfrac{46}{7}\right\}$

6. $|10x - 4| = 104$

(A) $\left\{6, -\dfrac{5}{2}\right\}$ (B) $\{-4\}$

(C) $\left\{\dfrac{54}{5}, -10\right\}$ (D) $\left\{-4, \dfrac{16}{3}\right\}$

ABSOLUTE VALUE #22

Solve the below equation.

7. $|2 - 3n| = 29$

(A) $\{10, -\frac{31}{4}\}$ (B) $\{-9, \frac{31}{3}\}$

(C) $\{7, -\frac{20}{3}\}$ (D) $\{2, 7\}$

8. $|-1 + 10v| = 89$

(A) $\{6, -\frac{22}{5}\}$ (B) $\{6\}$

(C) $\{9, -\frac{44}{5}\}$ (D) $\{9\}$

9. $|6n - 7| = 49$

(A) $\{\frac{28}{3}, -7\}$ (B) $\{7\}$

(C) $\{-\frac{5}{2}, 7\}$ (D) $\{7, -\frac{65}{9}\}$

Pre-Algebra Vol 2

ABSOLUTE VALUE #22

Solve the below equation.

10. $|3r - 4| = 11$

(A) $\left\{-7, \dfrac{19}{2}\right\}$ (B) $\left\{-5, \dfrac{27}{5}\right\}$

(C) $\left\{5, -\dfrac{7}{3}\right\}$ (D) $\{-5\}$

11. $|6r + 2| = 26$

(A) $\left\{-3, \dfrac{3}{7}\right\}$ (B) $\{-4, 6\}$

(C) $\{-4\}$ (D) $\left\{4, -\dfrac{14}{3}\right\}$

12. $|-8m - 5| = 45$

(A) $\left\{-\dfrac{25}{4}\right\}$ (B) $\left\{\dfrac{38}{5}, -10\right\}$

(C) $\left\{-\dfrac{3}{2}, -3\right\}$ (D) $\left\{-\dfrac{25}{4}, 5\right\}$

Solve the below equation.

13. |10n − 1| = 41

(A) $\left\{\dfrac{21}{5}, -4\right\}$ (B) $\left\{0, -\dfrac{14}{3}\right\}$

(C) $\left\{-\dfrac{7}{5}\right\}$ (D) $\left\{-\dfrac{7}{5}, 3\right\}$

14. |5 − 2x| = 17

(A) {−7, 9} (B) {8, −18}

(C) {−6, 11} (D) {8}

15. |k − 8| = 1

(A) {9} (B) {9, 7}

(C) $\left\{\dfrac{10}{3}, -6\right\}$ (D) $\left\{-4, \dfrac{19}{3}\right\}$

PROPORTIONS #23

Round your answer to the nearest whole number.

1. Mia enlarged the size of a triangle to a base of 6 cm which originally had a base of 2 cm and a height of 4 cm. What is the height of the enlarged figure?

   (A) 12 cm   (B) 15 cm   (C) 14 cm   (D) 18 cm

2. Lucy is drawing a picture with a height of 2 in. The original picture had width of 2 in and a length of 4 in. What is the length of lucy's drawing?

   (A) 2 in   (B) 1 in   (C) 4 in   (D) 3 in

3. A box of cherries cost $2. How many boxes of cherries can Tom buy for $20?

   (A) 10   (B) 40   (C) 11   (D) 9

4. One box of tomatoes cost $3. How many boxes can be purchased with $21?

   (A) 7   (B) 63   (C) 6   (D) 5

PROPORTIONS #23

Round your answer to the nearest whole number.

5. A bunch of turnips costs $2. How many bunches can be purchased with $14 ?

   (A) 6   (B) 4   (C) 7   (D) 28

6. A bunch of Lilly's costs $5. How many bunches can Julia buy for $45 ?

   (A) 32   (B) 9   (C) 7   (D) 5

7. Ethan bought a box of tennis balls for $7. How many boxes can Ethan purchase if he had $49 ?

   (A) 8   (B) 9   (C) 7   (D) 24

8. Tracy has a picture with dimensions of 18 in wide and 6 in tall. She is making a small replica of it with a height of 1 in. What will be the width of the replica she is creating ?

   (A) 3 in   (B) 5 in   (C) 6 in   (D) 108 in

PROPORTIONS #23

Round your answer to the nearest whole number.

9. Amy was planning a trip to Kuta falls and exchanges $72 to dubs the currency used at Kuta falls. How many dubs will she get in exchange if each dub equals to $3 ?

   (A) 15 Dubs   (B) 27 Dubs   (C) 20 Dubs   (D) 24 Dubs

10. Stella bought one package of Parsley for $3. How many packages of Parsley can Stella buy with $18 ?

    (A) 2   (B) 4   (C) 54   (D) 6

11. Rosy bought one package of watermelon for $2. How many packages of watermelon can Rosy buy with $18 ?

    (A) 11   (B) 9   (C) 36   (D) 10

12. One Tulip bulb costs $5 at the garden center. How many Tulip bulb can be purchased for $95?

    (A) 19   (B) 10   (C) 20   (D) 17

Round your answer to the nearest whole number.

13. The money used in Swiss is called Krots and the exchange rate is 7 Krots to $1. If Mat exchanges $20 how many Krots will he get?

    (A) 110 Krots   (B) 135 Krots   (C) 120 Krots   (D) 140 Krots

14. Bella enlarged the size of a picture to a width of 6 cm. What is the height, if its original dimensions are 2 in wide and 3 in tall?

    (A) 9 cm   (B) 18 cm   (C) 8 cm   (D) 2 cm

15. A bag of potatoes costs $3.50. How many bags can Lena purchase with $14?

    (A) 4   (B) 5   (C) 6   (D) 36

PROPORTIONS #24

Evaluate the below proportion.

1. $-\dfrac{5}{7} = \dfrac{m}{3}$

   (A) 9.5   (B) −2.14   (C) −7   (D) 1

2. $\dfrac{9}{b} = \dfrac{8}{4}$

   (A) 10   (B) 9.1   (C) 4.5   (D) −7.76

3. $\dfrac{5}{n} = -\dfrac{2}{8}$

   (A) −8   (B) −20   (C) 2.016   (D) −2.1

4. $\dfrac{9}{x} = \dfrac{10}{5}$

   (A) 4.5   (B) 3.7   (C) 4   (D) −2.2

PROPORTIONS #24

Evaluate the below proportion.

5. $\dfrac{8}{9} = \dfrac{x}{6}$

   (A) –8.4 (B) 5.33 (C) 7 (D) 3.2

6. $\dfrac{8}{9} = \dfrac{3}{p}$

   (A) 3 (B) –1 (C) 3.38 (D) –1.8

7. $-\dfrac{10}{b} = \dfrac{4}{6}$

   (A) 6.9 (B) 8.4 (C) –15 (D) –7

8. $-\dfrac{3}{9} = \dfrac{8}{r}$

   (A) –24 (B) –9.98 (C) –8.3 (D) –9

Evaluate the below proportion.

9. $\dfrac{2}{10} = -\dfrac{x}{8}$

   (A) –2   (B) 4.4   (C) 1.29   (D) –1.6

10. $\dfrac{9}{8} = -\dfrac{8}{n}$

   (A) –6   (B) –7.11   (C) 5   (D) –8.854

11. $\dfrac{x}{7} = -\dfrac{8}{4}$

   (A) 2.5   (B) 1.5   (C) –14   (D) –5.3

12. $\dfrac{9}{5} = -\dfrac{4}{k}$

   (A) –2.22   (B) –7.2   (C) 2.22   (D) –7

PROPORTIONS #24

Evaluate the below proportion.

13. $\dfrac{k}{7} = \dfrac{4}{5}$

(A) −1.8 (B) 4.5 (C) 5.6 (D) 7.6

14. $\dfrac{2}{4} = \dfrac{n}{3}$

(A) 1.5 (B) 9.4 (C) 9.2 (D) −5

15. $\dfrac{r}{5} = -\dfrac{2}{4}$

(A) −1 (B) −2.5 (C) −5.6 (D) 4.2

PERCENT DISCOUNT #25

1. Jack purchased a video game, originally priced at $52.50. He has a coupon for 57% discount. How much did Jack paid to the store for the game ?

   (A) $22.575   (B) $22.557   (C) $25.75   (D) $25.257

2. A witting desk is priced at $249.50 and the discount offered on it is 55%, calculate the selling price.

   (A) $112.27   (B) $137.23   (C) $112.28   (D) $386.73

3. David has a 5% discount coupon for any purchases at store A. All hand bags are priced at $99.99. What is the selling price of one bag ?

   (A) $79.99   (B) $5.00   (C) $104.99   (D) $94.99

4. A dress is priced at $134.99 and the store is offering a 40% discount. What is the selling price of the dress ?

   (A) $188.99   (B) $114.74   (C) $121.49   (D) $80.99

PERCENT DISCOUNT #25

5. Cathy purchased a fish tank for $149.95. Find the selling price of the fish tank if a discount of 50% is offered on it ?

   (A) $224.92   (B) $74.92   (C) $74.98   (D) $164.94

6. An electronic store in town A sells computers at $850.99. A store is offering a 40% discount today. Find the selling price of the computer ?

   (A) $510.594   (B) $511.54   (C) $510.509   (D) $510.67

7. Brand X of a vaccum cleaner is sold at a price of $109.95. A store is offering a 40% discount today. Find the selling price of the vaccum cleaner.

   (A) $131.94   (B) $93.46   (C) $65.97   (D) $43.98

8. The price of a refrigerator is $459.99. The store is offering a discount of 40% on it. What is the selling price of refrigerator ?

   (A) $275.99   (B) $643.99   (C) $184.00   (D) $528.99

PERCENT DISCOUNT #25

9. Sports stores is selling skate board at $269.99. Ben has a coupon for 40% discount. Find the selling price of the skate board for ben ?

   (A) $377.99   (B) $161.99   (C) $108.00   (D) $310.49

10. A mini car is on sale with a discount of 10% which is originally priced at $25,300. Find the selling price of mini car ?

    (A) $20.240.00 (B) $20,095.00 (C) $2,530.00   (D) $22,770.00

11. A mini van is on sale with a discount of 22% which is originally priced at $45,000.00. Find the selling price of mini van ?

    (A) $35,100.00 (B) $51,750.00 (C) $49,500.00 (D) $9,900.00

12. The original price of the furniture is $850.00 and a discount of 50% is offered on it. What is the selling price of the furniture ?

    (A) $765.00   (B) $680.00   (C) $425.00   (D) $1.020.00

13. A pendant is originally priced at $83.50. The store is offering 20% discount on it. What is the selling price of it ?

   (A) $16.70   (B) $87.67   (C) $75.15   (D) $66.80

14. What is the selling price of a video game console that is priced $499.50 and when a 60% discount is offered on it ?

   (A) $474.52   (B) $799.20   (C) $299.70   (D) $199.80

15. A backpack is priced at $10.00. The store is offering a discount of 35% on it. Find the selling price of it ?

   (A) $13.50   (B) $10.50   (C) $3.50   (D) $6.50

Pre-Algebra Vol 2

PERCENT MARKUP #26

1. Amy purchased a shirt at $35.5 and sells it at a 20% price increase. Find the selling price of the shirt.

   (A) $42.60   (B) $39.05   (C) $7.10   (D) $28.40

2. Beth purchased a puzzle set at $2.50 and sells it at a 39% price increase. Find the selling price of the puzzle set.

   (A) $2.75   (B) $3.48   (C) $0.98   (D) $1.52

3. Frank purchased a box of apples at $44.95 and sells it at a 65% price increase. Find the selling price of the box of apples.

   (A) $40.46   (B) $74.17   (C) $15.73   (D) $29.22

4. Julia purchased a game for $39.99 and sold at 25% price hike. What is the selling price of the game?

   (A) $29.99   (B) $49.99   (C) $31.99   (D) $10.00

PERCENT MARKUP #26

5. Zoya purchased an office furniture at $13,000.00 and sells it at a 45% price hike. Find the selling price of the office furniture?

   (A) $18,850.00   (B) $5,850.00   (C) $17,907.50   (D) $7,150.00

6. Lilly purchased a box of cupcakes at $3.50 and sells it at a 20% price increase. What is the selling price of the box of cupcakes?

   (A) $4.02   (B) $4.20   (C) $2.80   (D) $0.70

7. Ila purchased a shirt at $50.00 and sells it at a 65% price increase. Find the selling price of the shirt.

   (A) $32.50   (B) $17.50   (C) $82.50   (D) $60.00

8. Dany purchased a hover board at $219.50 and sells it at a 36% price increase. Find the selling price of the hover board.

   (A) $140.48   (B) $175.60   (C) $298.52   (D) $79.02

PERCENT MARKUP #26

9. Hazel purchased a book at $6.95 and sells it at a 35% price increase. Find the selling price of the book.

   (A) $4.52   (B) $9.38   (C) $2.43   (D) $6.60

10. Clara purchased an air hockey game at $119.50 and sells it at a 15% price increase. Find the selling price of the air hockey game.

    (A) $137.43   (B) $137.05   (C) $101.58   (D) $17.93

11. Owen purchased a digital camera at $140.00 and sells it at a 80% price increase. Find the selling price of the digital camera.

    (A) $252.00   (B) $168.00   (C) $126.00   (D) $112.00

12. Dean purchased a pair of plants at $20.50 and sells it at a 84% price increase. Find the selling price of the pair of plants.

    (A) $18.45   (B) $3.28   (C) $17.22   (D) $37.72

13. Hank purchased a theme park ticket at $149.95 and sells it at a 90% price increase. Find the selling price of the theme park ticket.

   (A) $284.99   (B) $157.45   (C) $14.99   (D) $284.91

14. Flora purchased a roller skates at $100.00 and sells it at a 90% price increase. Find the selling price of the roller skates.

   (A) $190.00   (B) $10.00   (C) $90.00   (D) $85.00

15. Violet purchased a cell phone X at $799.99 and sells it at a 13% price increase. Find the selling price of the cell phone X.

   (A) $104.00   (B) $959.99   (C) $919.99   (D) $903.99

PERCENT TAX #27

1. Joya purchased an X - box game which is priced at $43.99. If a tax of 4% need to added to the price, what is the selling price of it?

   (A) $37.39   (B) $42.23   (C) $1.76   (D) $45.75

2. The price of a parrot at a local pet shop is $119.95. Blake pays a 3% tax on the price. Find the selling price.

   (A) $3.60   (B) $123.55   (C) $116.35   (D) $143.94

3. Calculate the sale price of an ipad when 6% of tax is added to its original price at $300.00

   (A) $18.99   (B) $285.00   (C) $282.00   (D) $318.00

4. The cost of a telescope is $149.95. Calculate the selling price when 3% of tax is added to it.

   (A) $142.45   (B) $4.50   (C) $164.94   (D) $154.45

PERCENT TAX #27

5. Jasper purchased a scooter for $35,000.00. He was charged 2% tax on the price. Calculate the sale price of the scooter ?

   (A) $28,000.00 (B) $700.00   (C) $35,700.00 (D) $34,300.00

6. George purchased a chocolate bar for $2.20. A 2% is added to his bill. Find the amount George paid ?

   (A) $0.04   (B) $2.16   (C) $1.98   (D) $2.24

7. The original price of a pen is $2.50, calculate the selling price of it with a 3% tax added to it.

   (A) $2.12   (B) $2.58   (C) $0.07   (D) $2.42

8. The original price of a box of pencils is $78.95, calculate the selling price of it with a 2% tax added to it.

   (A) $67.11   (B) $77.37   (C) $1.58   (D) $80.53

PERCENT TAX #27

9. David purchased a game for $13.50. Find the selling price with a 5% tax added to it ?

   (A) $14.18   (B) $15.52   (C) $16.20   (D) $12.82

10. Ada purchased a hat for $18.99. Find the selling price with a 3% tax added to it ?

    (A) $19.94   (B) $19.56   (C) $18.42   (D) $22.79

11. A bag of chips cost $3.99 and a 3% of tax is added to it, calculate the selling price ?

    (A) $4.11   (B) $4.79   (C) $0.12   (D) $3.87

12. A cost of a chair is $249.50 and a 5% of tax is added to it, calculate the selling price ?

    (A) $237.02   (B) $274.45   (C) $261.98   (D) $12.48

PERCENT TAX #27

13. Eli purchased a patio furniture set for $194.95. Find out the selling price if 4% tax is added to it.

    (A) $187.15   (B) $202.75   (C) $214.44   (D) $7.80

14. Find the selling price of a Jacket that is priced $67.00 and a tax of 6% is charged on it ?

    (A) $80.40   (B) $71.02   (C) $62.98   (D) $4.02

15. The original price of a basketball is $64.99 and a 2% of tax is added to it. Calculate the selling price.

    (A) $1.30   (B) $63.69   (C) $74.74   (D) $66.29

Pre-Algebra Vol 2

PERCENT CHANGE #28

1. Find the percentage change from 91 to 39.

   (A) 57.1% decrease
   (B) 57.1% increase
   (C) 233.3% decrease
   (D) 52% increase

2. Find the percentage change from 74 to 17

   (A) 335.3% increase
   (B) 335.3% decrease
   (C) 77% decrease
   (D) 435.3% decrease

3. Find the percentage change from 30.2 to 21

   (A) 69.5% decrease
   (B) 9.2% decrease
   (C) 9.2% increase
   (D) 30.5% decrease

4. Find the percentage change from 52 to 98.4

   (A) 58.9% increase
   (B) 47.2% decrease
   (C) 89.2% decrease
   (D) 89.2% increase

PERCENT CHANGE #28

5. Find the percentage change from 71 to 4.3

   (A) 66.7% increase   (B) 1651.2% decrease
   (C) 93.9% decrease   (D) 6.1% decrease

6. Find the percentage change from 37 to 22

   (A) 40.5% decrease   (B) 68.2% decrease
   (C) 68.2% increase   (D) 59.5% decrease

7. Find the percentage change from 81 to 98

   (A) 21% increase   (B) 121% increase
   (C) 17.3% increase   (D) 17% increase

8. Find the percentage change from 41 to 33

   (A) 24.2% increase   (B) 8% increase
   (C) 19.5% decrease   (D) 8% decrease

PERCENT CHANGE #28

9. Find the percentage change from 91 to 11

(A) 827.3% decrease (B) 36.4% decrease
(C) 87.9% decrease (D) 12.1% decrease

10. Find the percentage change from 41 to 29

(A) 29.3% decrease (B) 41.4% increase
(C) 29.3% increase (D) 41.4% decrease

11. Find the percentage change from 23 to 17

(A) 65.7% increase (B) 6% increase
(C) 73.9% decrease (D) 26.1% decrease

12. Find the percentage change from 60 to 58

(A) 3.3% increase (B) 3.3% decrease
(C) 96.7% decrease (D) 2% decrease

13. Find the percentage change from 93 to 67

(A) 95.5% increase (B) 28% decrease
(C) 72% decrease (D) 26% increase

14. Find the percentage change from 64 to 66

(A) 3.1% decrease (B) 103.1% increase
(C) 2% decrease (D) 3.1% increase

15. Find the percentage change from 97 to 57

(A) 40% increase (B) 41.2% increase
(C) 40% decrease (D) 41.2% decrease

Pre-Algebra Vol 2

RADICALS #29

Find the square root of the below.

1. $\boxed{\sqrt{\dfrac{100}{169}}}$

   (A) $\dfrac{5}{6}$ (B) $\dfrac{2}{7}$ (C) $\dfrac{10}{13}$ (D) $\dfrac{4}{7}$

2. $\boxed{-\sqrt{\dfrac{169}{144}}}$

   (A) $-2\dfrac{1}{3}$ (B) $-\dfrac{11}{13}$ (C) $-1\dfrac{1}{12}$ (D) $-\dfrac{4}{11}$

3. $\boxed{\sqrt{\dfrac{49}{4}}}$

   (A) $\dfrac{7}{9}$ (B) $3\dfrac{1}{2}$ (C) $\dfrac{3}{11}$ (D) $\dfrac{7}{13}$

4. $\boxed{-\sqrt{\dfrac{16}{121}}}$

   (A) $-\dfrac{2}{3}$ (B) $-\dfrac{7}{9}$ (C) $-\dfrac{4}{11}$ (D) $-\dfrac{11}{14}$

RADICALS #29

Find the square root of the below.

5. $\sqrt{\dfrac{1}{64}}$

(A) $\dfrac{1}{8}$ (B) $\dfrac{2}{11}$ (C) $\dfrac{1}{9}$ (D) $\dfrac{8}{13}$

6. $\sqrt{\dfrac{36}{144}}$

(A) $\dfrac{3}{4}$ (B) $1\dfrac{5}{8}$ (C) $\dfrac{1}{2}$ (D) $\dfrac{3}{13}$

7. $\sqrt{\dfrac{49}{100}}$

(A) $\dfrac{9}{10}$ (B) $\dfrac{7}{10}$ (C) $\dfrac{1}{3}$ (D) $\dfrac{5}{9}$

8. $\sqrt{\dfrac{64}{9}}$

(A) $\dfrac{6}{11}$ (B) $\dfrac{1}{2}$ (C) $2\dfrac{2}{3}$ (D) $\dfrac{9}{13}$

RADICALS #29

Find the square root of the below.

9. $\boxed{-\sqrt{\dfrac{16}{169}}}$

(A) $-\dfrac{1}{3}$ (B) $-\dfrac{12}{13}$ (C) $-\dfrac{4}{13}$ (D) -2

10. $\boxed{\sqrt{\dfrac{1}{100}}}$

(A) $\dfrac{3}{5}$ (B) $\dfrac{1}{8}$ (C) $\dfrac{1}{5}$ (D) $\dfrac{1}{10}$

11. $\boxed{-\sqrt{\dfrac{9}{49}}}$

(A) $-\dfrac{1}{3}$ (B) $-\dfrac{1}{8}$ (C) $-\dfrac{3}{4}$ (D) $-\dfrac{3}{7}$

12. $\boxed{-\sqrt{\dfrac{4}{36}}}$

(A) $-\dfrac{7}{10}$ (B) $-\dfrac{6}{13}$ (C) $-\dfrac{1}{3}$ (D) $-\dfrac{9}{10}$

Find the square root of the below.

13. $-\sqrt{\dfrac{16}{4}}$

(A) $-\dfrac{6}{13}$ (B) $-\dfrac{5}{14}$ (C) -2 (D) $-\dfrac{7}{8}$

14. $\sqrt{\dfrac{121}{4}}$

(A) $5\dfrac{1}{2}$ (B) $\dfrac{1}{12}$ (C) $\dfrac{2}{5}$ (D) $\dfrac{8}{13}$

15. $\sqrt{\dfrac{25}{81}}$

(A) $\dfrac{2}{9}$ (B) $\dfrac{5}{9}$ (C) $\dfrac{5}{6}$ (D) $\dfrac{6}{7}$

RADICALS #30

Simplify the below radicals.

1. $\boxed{-2\sqrt{3} - 2\sqrt{5} + 3\sqrt{27}}$

   (A) $5\sqrt{3} - 6\sqrt{5}$ (B) $7\sqrt{3} - 2\sqrt{5}$

   (C) $7\sqrt{3} - 4\sqrt{5}$ (D) $7\sqrt{3} - 6\sqrt{5}$

2. $\boxed{-3\sqrt{24} + 2\sqrt{6} + 3\sqrt{24}}$

   (A) $8\sqrt{6}$ (B) $-4\sqrt{6}$

   (C) $14\sqrt{6}$ (D) $2\sqrt{6}$

3. $\boxed{-3\sqrt{2} - \sqrt{45} - 2\sqrt{45}}$

   (A) $-6\sqrt{2} - 9\sqrt{5}$ (B) $-3\sqrt{2} - 9\sqrt{5}$

   (C) $-9\sqrt{2} - 15\sqrt{5}$ (D) $-6\sqrt{2} - 15\sqrt{5}$

4. $\boxed{-\sqrt{2} - \sqrt{5} - 2\sqrt{5}}$

   (A) $-\sqrt{2} - 8\sqrt{5}$ (B) $-\sqrt{2} - 3\sqrt{5}$

   (C) $-\sqrt{2} - 5\sqrt{5}$ (D) $-\sqrt{2} - 7\sqrt{5}$

RADICALS #30

Simplify the below radicals.

5. $\boxed{-2\sqrt{24} - 2\sqrt{54} + 3\sqrt{6}}$

   (A) $-4\sqrt{6}$ (B) $2\sqrt{6}$

   (C) $-7\sqrt{6}$ (D) $-\sqrt{5}$

6. $\boxed{2\sqrt{20} - 2\sqrt{2} - 2\sqrt{5}}$

   (A) $-4\sqrt{2}$ (B) $2\sqrt{5} - 2\sqrt{2}$

   (C) $2\sqrt{5} - 4\sqrt{2}$ (D) $-6\sqrt{2}$

7. $\boxed{-3\sqrt{5} - \sqrt{54} + 3\sqrt{20}}$

   (A) $9\sqrt{5} - 3\sqrt{6}$ (B) $3\sqrt{5} - 3\sqrt{6}$

   (C) $12\sqrt{5} - 3\sqrt{6}$ (D) $15\sqrt{5} - 3\sqrt{6}$

8. $\boxed{-\sqrt{27} - 3\sqrt{20} + 2\sqrt{3}}$

   (A) $-5\sqrt{3} - 6\sqrt{5}$ (B) $-2\sqrt{3} - 6\sqrt{5}$

   (C) $\sqrt{3} - 6\sqrt{5}$ (D) $-\sqrt{3} - 6\sqrt{5}$

Simplify the below radicals.

9. $\boxed{-2\sqrt{6} + 3\sqrt{20} - 2\sqrt{45}}$

(A) $-4\sqrt{6} + 12\sqrt{5}$ (B) $-2\sqrt{6}$
(C) $-2\sqrt{6} + 12\sqrt{5}$ (D) $-2\sqrt{6} + 6\sqrt{5}$

10. $\boxed{-2\sqrt{18} - \sqrt{45} - 3\sqrt{18}}$

(A) $-21\sqrt{2} - 3\sqrt{5}$ (B) $-21\sqrt{2} - 9\sqrt{5}$
(C) $-15\sqrt{2} - 3\sqrt{5}$ (D) $-21\sqrt{2} - 6\sqrt{5}$

11. $\boxed{-2\sqrt{3} - 2\sqrt{3} + 3\sqrt{27}}$

(A) $12\sqrt{3}$ (B) $5\sqrt{3}$
(C) $14\sqrt{3}$ (D) $10\sqrt{3}$

12. $\boxed{-3\sqrt{45} - 2\sqrt{20} + 2\sqrt{3}}$

(A) $-25\sqrt{5} + 2\sqrt{3}$ (B) $-13\sqrt{5} + 2\sqrt{3}$
(C) $-21\sqrt{5} + 2\sqrt{3}$ (D) $-17\sqrt{5} + 2\sqrt{3}$

Simplify the below radicals.

13. $-3\sqrt{54} + 2\sqrt{24} + 2\sqrt{8}$

   (A) $-10\sqrt{6} + 8\sqrt{2}$ (B) $-\sqrt{6} + 8\sqrt{2}$
   (C) $-5\sqrt{6} + 4\sqrt{2}$ (D) $-5\sqrt{6} + 8\sqrt{2}$

14. $-3\sqrt{18} - \sqrt{6} - 2\sqrt{6}$

   (A) $-18\sqrt{2} - 6\sqrt{6}$ (B) $-9\sqrt{2} - 3\sqrt{6}$
   (C) $-9\sqrt{2} - 5\sqrt{6}$ (D) $-18\sqrt{2} - 5\sqrt{6}$

15. $2\sqrt{24} + 2\sqrt{54} - 2\sqrt{2}$

   (A) $10\sqrt{6} - 4\sqrt{2}$ (B) $16\sqrt{6} - 6\sqrt{2}$
   (C) $10\sqrt{6} - 6\sqrt{2}$ (D) $10\sqrt{6} - 2\sqrt{2}$

RADICALS #31

Simplify the below radicals.

1. $\sqrt{3}(\sqrt{6}+\sqrt{2})$

   (A) $3\sqrt{2}+\sqrt{6}$ (B) 13
   (C) $-3\sqrt{3}+5$ (D) $-3\sqrt{30}+2$

2. $4\sqrt{15}(\sqrt{3}+5\sqrt{5})$

   (A) $12\sqrt{5}+100\sqrt{3}$ (B) $\sqrt{30}+2$
   (C) $4\sqrt{3}+6$ (D) $5\sqrt{30}-12\sqrt{5}$

3. $5\sqrt{6}(5+\sqrt{3})$

   (A) $\sqrt{30}+4$ (B) $25\sqrt{6}+15\sqrt{2}$
   (C) $4\sqrt{5}+2$ (D) $2\sqrt{2}+3$

4. $-\sqrt{10}(2+\sqrt{2})$

   (A) $4\sqrt{5}+\sqrt{30}$ (B) $9\sqrt{3}$
   (C) $-2\sqrt{10}-2\sqrt{5}$ (D) $8+4\sqrt{30}$

Simplify the below radicals.

5. $\sqrt{15}(-4\sqrt{10}+3)$

(A) $-22\sqrt{3}$ (B) $2\sqrt{3}+3$
(C) $2\sqrt{5}+4$ (D) $-20\sqrt{6}+3\sqrt{15}$

6. $\sqrt{10}(-4\sqrt{2}+4)$

(A) $\sqrt{30}+3$ (B) $16\sqrt{2}$
(C) 13 (D) $-8\sqrt{5}+4\sqrt{10}$

7. $\sqrt{15}(\sqrt{5}+4)$

(A) $2\sqrt{3}-8\sqrt{5}$ (B) $-30+25\sqrt{3}$
(C) $5\sqrt{3}+4\sqrt{15}$ (D) $\sqrt{30}+3$

8. $4\sqrt{6}(\sqrt{3}+4)$

(A) $2\sqrt{5}+4$ (B) $2\sqrt{3}+4$
(C) $3\sqrt{3}-8\sqrt{5}$ (D) $12\sqrt{2}+16\sqrt{6}$

Simplify the below radicals.

9. $\boxed{-2\sqrt{3}(\sqrt{6} + 4)}$

   (A) 0
   (B) $-6\sqrt{2} - 8\sqrt{3}$
   (C) 8
   (D) $5\sqrt{5} + 16$

10. $\boxed{\sqrt{15}(4\sqrt{10} + 5)}$

    (A) $20\sqrt{6} + 5\sqrt{15}$
    (B) $2\sqrt{2} + 4$
    (C) $-2\sqrt{5} + 5$
    (D) $\sqrt{30} + 2$

11. $\boxed{\sqrt{15}(2 - 5\sqrt{6})}$

    (A) $20\sqrt{3} + 5$
    (B) $5\sqrt{2} + 8$
    (C) $5\sqrt{2} + 15\sqrt{5}$
    (D) $2\sqrt{15} - 15\sqrt{10}$

12. $\boxed{\sqrt{5}(3 + 3\sqrt{3})}$

    (A) $-10\sqrt{2} + 2$
    (B) 23
    (C) $3\sqrt{5} + 3\sqrt{15}$
    (D) $8\sqrt{5}$

RADICALS #31

Simplify the below radicals.

13. $\boxed{-\sqrt{2}(\sqrt{2}+5)}$

(A) $-2 - 5\sqrt{2}$
(B) $5\sqrt{2} + 8$
(C) $20\sqrt{5} + 3$
(D) $2\sqrt{3} + \sqrt{30}$

14. $\boxed{\sqrt{10}(-\sqrt{2}+4)}$

(A) $-2\sqrt{5} + 4\sqrt{10}$
(B) 9
(C) $4\sqrt{2} + 4$
(D) $4\sqrt{3} + 2$

15. $\boxed{\sqrt{6}(\sqrt{2}+3\sqrt{5})}$

(A) $25\sqrt{2} + 2$
(B) $-15\sqrt{2} + 2$
(C) $8\sqrt{2}$
(D) $2\sqrt{3} + 3\sqrt{30}$

RADICALS #32

Simplify the below radicals.

1. $$\dfrac{\sqrt{2}}{5+3\sqrt{5}}$$

   (A) $\dfrac{-3\sqrt{3}+9}{4}$ (B) $\dfrac{5-3\sqrt{2}}{5}$

   (C) $\dfrac{-5\sqrt{2}+3\sqrt{10}}{20}$ (D) $-6+3\sqrt{5}$

2. $$\dfrac{\sqrt{5}}{3-\sqrt{2}}$$

   (A) $\dfrac{3\sqrt{5}+\sqrt{10}}{7}$ (B) $\dfrac{6\sqrt{5}+2\sqrt{2}}{43}$

   (C) $\dfrac{-3-5\sqrt{5}}{58}$ (D) $\dfrac{6+2\sqrt{3}}{3}$

3. $$\dfrac{4}{2-4\sqrt{5}}$$

   (A) $\dfrac{-\sqrt{2}-2\sqrt{3}}{10}$ (B) $\dfrac{1-2\sqrt{5}}{2}$

   (C) $\dfrac{-2-4\sqrt{5}}{19}$ (D) $\dfrac{-5-5\sqrt{5}}{16}$

Simplify the below radicals.

4. $\dfrac{5}{4\sqrt{2}+2\sqrt{3}}$

(A) $\dfrac{2\sqrt{2}-\sqrt{3}}{2}$ (B) $\dfrac{-12-8\sqrt{3}}{3}$

(C) $\dfrac{-12-3\sqrt{2}}{14}$ (D) $\dfrac{-12+16\sqrt{5}}{71}$

5. $\dfrac{4}{-5+\sqrt{3}}$

(A) $\dfrac{-5+\sqrt{3}}{4}$ (B) $\dfrac{-10-5\sqrt{2}}{2}$

(C) $\dfrac{-1+\sqrt{5}}{4}$ (D) $\dfrac{-10-2\sqrt{3}}{11}$

6. $\dfrac{4}{4+2\sqrt{2}}$

(A) $2-\sqrt{2}$ (B) $\dfrac{4\sqrt{5}-5}{11}$

(C) $\dfrac{-3\sqrt{2}-\sqrt{10}}{4}$ (D) $\dfrac{-2-\sqrt{2}}{2}$

Simplify the below radicals.

7. $\dfrac{2}{4+\sqrt{5}}$

(A) $\dfrac{4+\sqrt{5}}{2}$

(B) $\dfrac{8-2\sqrt{5}}{11}$

(C) $\dfrac{3\sqrt{5}+\sqrt{3}}{21}$

(D) $\dfrac{6-2\sqrt{2}}{7}$

8. $\dfrac{5}{4-\sqrt{2}}$

(A) $-1+\sqrt{2}$

(B) $\dfrac{2+2\sqrt{3}}{3}$

(C) $\dfrac{20+5\sqrt{2}}{14}$

(D) $\dfrac{-5\sqrt{3}-\sqrt{6}}{3}$

9. $\dfrac{5}{-3-\sqrt{2}}$

(A) $\dfrac{-12-9\sqrt{3}}{11}$

(B) $\dfrac{15+9\sqrt{2}}{7}$

(C) $\dfrac{-15+5\sqrt{2}}{7}$

(D) $4+2\sqrt{3}$

Simplify the below radicals.

10. $\boxed{\dfrac{2}{-2-\sqrt{3}}}$

(A) $\dfrac{1+\sqrt{3}}{5}$

(B) $4 - 2\sqrt{2}$

(C) $-4 + 2\sqrt{3}$

(D) $\dfrac{2+\sqrt{3}}{2}$

11. $\boxed{-\dfrac{2}{3-2\sqrt{2}}}$

(A) $-5 + 3\sqrt{3}$

(B) $\sqrt{5} - 1$

(C) $-6 - 4\sqrt{2}$

(D) $\dfrac{-3 + 2\sqrt{2}}{2}$

12. $\boxed{\dfrac{3}{5\sqrt{2} + 3}}$

(A) $\dfrac{10 - 5\sqrt{2}}{2}$

(B) $\dfrac{-3 + \sqrt{5}}{4}$

(C) $\dfrac{15\sqrt{2} - 9}{41}$

(D) $\dfrac{-2 - 3\sqrt{6}}{25}$

RADICALS #32

Simplify the below radicals.

13. $\dfrac{5}{4 - 4\sqrt{2}}$

(A) $\dfrac{4 - 4\sqrt{2}}{5}$

(B) $\dfrac{-5 - 5\sqrt{2}}{4}$

(C) $\dfrac{3 - 3\sqrt{5}}{8}$

(D) $-10 + 5\sqrt{5}$

14. $\dfrac{4}{3\sqrt{3} - \sqrt{2}}$

(A) $\dfrac{-6 - 8\sqrt{3}}{39}$

(B) $\dfrac{-16 - 4\sqrt{3}}{13}$

(C) $\dfrac{3 + \sqrt{3}}{2}$

(D) $\dfrac{12\sqrt{3} + 4\sqrt{2}}{25}$

15. $\dfrac{3}{4 - 3\sqrt{5}}$

(A) $-2\sqrt{3} + 2\sqrt{5}$

(B) $\dfrac{-15 + 5\sqrt{2}}{7}$

(C) $\dfrac{-9 + 15\sqrt{2}}{41}$

(D) $\dfrac{-12 - 9\sqrt{5}}{29}$

Evaluate the below equation and find the value of the variable.

1. $6 + \sqrt{11 - r} = 7$

   (A) 10 (B) 6 (C) −6 (D) −1, 10

2. $21 = 3\sqrt{7n}$

   (A) −4 (B) 7 (C) 4 (D) 7, 4

3. $-5\sqrt{\dfrac{m}{8}} = -40$

   (A) −8, 2 (B) 2 (C) 6 (D) 512

4. $-10\sqrt{1 - v} = -10$

   (A) 5 (B) 0 (C) 8, 3 (D) 3

Evaluate the below equation and find the value of the variable.

5. $3 = \sqrt{\dfrac{x}{7}}$

(A) 2, -1   (B) -1   (C) 63   (D) 8

6. $9 = \sqrt{8m + 9}$

(A) 9, -9   (B) 5   (C) -4   (D) 9

7. $5 = \sqrt{26x - 1}$

(A) 1, -1   (B) 1, -8   (C) 4   (D) 1

8. $8 = 3 + \sqrt{8a + 1}$

(A) 10   (B) 2   (C) 4   (D) 3

Evaluate the below equation and find the value of the variable.

9. $10 = \sqrt{19x + 5}$

(A) −5   (B) 5   (C) 5, −5   (D) 4, −5

10. $\sqrt{3k + 54} = 9$

(A) −9, 2   (B) −2   (C) −9   (D) 9

11. $4 = \sqrt{k - 7}$

(A) 23   (B) 8   (C) −23   (D) 9

12. $0 = \sqrt{n + 6}$

(A) 6   (B) −6   (C) 2, −6   (D) 7, 6

Evaluate the below equation and find the value of the variable.

13. $8 = \sqrt{r+4} + 1$

   (A) −8, −4 (B) 8, −4 (C) 45 (D) −8

14. $6 + \sqrt{-4n} = 6$

   (A) 2 (B) 7 (C) −9 (D) 0

15. $8 = 2 + \sqrt{\dfrac{n}{6}}$

   (A) 215, −5 (B) 216 (C) 216, 3 (D) 216, −8

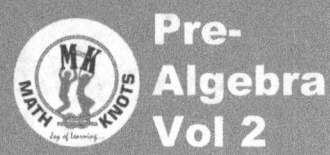

RADICALS #34

Evaluate the below equation and find the value of the variable.

1. $\sqrt{108b^3}$

   (A) $6b\sqrt{3b}$ (B) $5\sqrt{5b}$ (C) $10\sqrt{b}$ (D) $2b^2\sqrt{3}$

2. $\sqrt{128p^4}$

   (A) $6p\sqrt{5p}$ (B) $8p^2\sqrt{2}$ (C) $4p^2\sqrt{3}$ (D) $7p\sqrt{3p}$

3. $\sqrt{200r^2}$

   (A) $16r^2\sqrt{2}$ (B) $10r\sqrt{2}$ (C) $8\sqrt{2r}$ (D) $7\sqrt{2r}$

4. $\sqrt{32m^3}$

   (A) $3\sqrt{3m}$ (B) $3m^2\sqrt{5}$ (C) $4m\sqrt{2m}$ (D) $5\sqrt{2m}$

RADICALS #34

Evaluate the below equation and find the value of the variable.

5. $\sqrt{8m^2}$

(A) $2m\sqrt{5m}$ (B) $2\sqrt{7m}$ (C) $2m\sqrt{6}$ (D) $2m\sqrt{2}$

6. $\sqrt{294k^2}$

(A) $4k^2\sqrt{2}$ (B) $6k\sqrt{6k}$ (C) $8k\sqrt{6k}$ (D) $7k\sqrt{6}$

7. $\sqrt{80p^3}$

(A) $4\sqrt{2p}$ (B) $3p^2\sqrt{6}$ (C) $4p\sqrt{5p}$ (D) $2p\sqrt{5}$

8. $\sqrt{112v^3}$

(A) $8\sqrt{2v}$ (B) $7v\sqrt{6v}$ (C) $4v^2\sqrt{2}$ (D) $4v\sqrt{7v}$

Evaluate the below equation and find the value of the variable.

9. $\sqrt{216p^3}$

(A) $6\sqrt{p}$ (B) $6p\sqrt{6p}$ (C) $3\sqrt{6p}$ (D) $6p\sqrt{7p}$

10. $\sqrt{64v^4}$

(A) $8v^2$ (B) $8v\sqrt{2v}$ (C) $2v\sqrt{2v}$ (D) $7v\sqrt{5v}$

11. $\sqrt{512v^2}$

(A) $10v^2$ (B) $7\sqrt{6v}$ (C) $6v\sqrt{2v}$ (D) $16v\sqrt{2}$

12. $\sqrt{256r^4}$

(A) $6r\sqrt{6r}$ (B) $16r^2$ (C) $14r^2\sqrt{2}$ (D) $8r\sqrt{6r}$

Evaluate the below equation and find the value of the variable.

13. $\sqrt{64p^2}$

(A) $2p^2\sqrt{7}$ (B) $8p$ (C) $7p\sqrt{2p}$ (D) $7p^2\sqrt{3}$

14. $\sqrt{112n^2}$

(A) $5n^2\sqrt{7}$ (B) $2n\sqrt{6n}$ (C) $7n\sqrt{3}$ (D) $4n\sqrt{7}$

15. $\sqrt{128b^3}$

(A) $4b\sqrt{2}$ (B) $16b^2$ (C) $8b\sqrt{2b}$ (D) $2b\sqrt{2}$

Pre-Algebra Vol 2

Triangles #35

Find the area of the below triangles :

1)

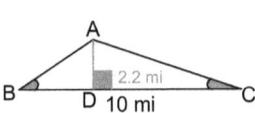

A) 11 mi²   B) 22 mi²
C) 17.1 mi²   D) 5.5 mi²

2)

A) 46 ft²   B) 91.95 ft²
C) 83.85 ft²   D) 167.7 ft²

3)

A) 34.7 km²   B) 138.7 km²
C) 64.45 km²   D) 69.35 km²

4)

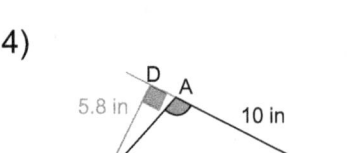

A) 58 in²   B) 29 in²
C) 36.8 in²   D) 14.5 in²

5)

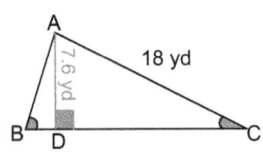

A) 136.8 yd²   B) 70.6 yd²
C) 34.2 yd²   D) 68.4 yd²

6)

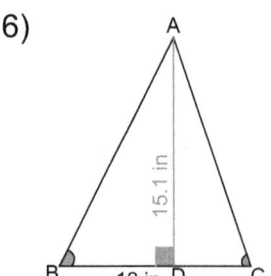

A) 98.15 in²   B) 196.3 in²
C) 49.1 in²   D) 88.85 in²

Triangles #35

Find the area of the below triangles :

7)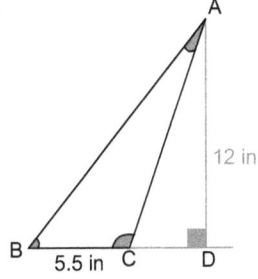

A) 37.5 in² B) 66 in²
C) 33 in²   D) 16.5 in²

8)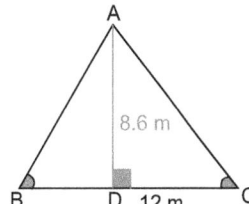

A) 25.8 m²   B) 42.2 m²
C) 103.2 m² D) 51.6 m²

9)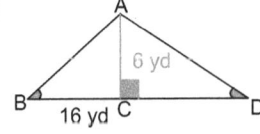

A) 24 yd²   B) 48 yd²
C) 40.3 yd² D) 96 yd²

10)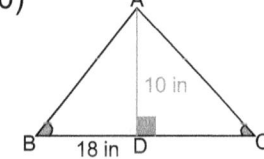

A) 90 in² B) 180 in²
C) 45 in² D) 80 in²

11)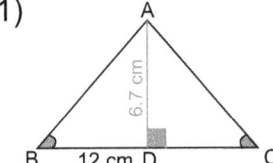

A) 20.1 cm² B) 50 cm²
C) 80.4 cm² D) 40.2 cm²

12)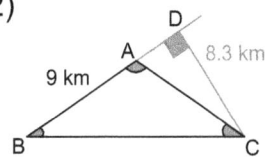

A) 74.7 km²   B) 28.95 km²
C) 37.35 km² D) 18.7 km²

Triangles #35

Find the area of the below triangles :

13)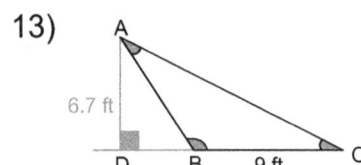

A) 15.1 ft² B) 32.65 ft²
C) 30.15 ft² D) 65.3 ft²

14)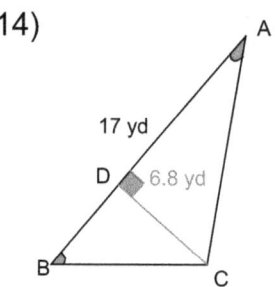

A) 115.6 yd² B) 64.5 yd²
C) 57.8 yd² D) 129 yd²

15)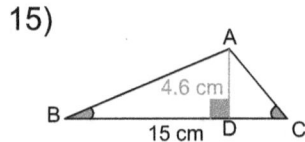

A) 34.5 cm² B) 47.4 cm²
C) 38.6 cm² D) 17.3 cm²

Identify the below quadrilateral and find its area :

1)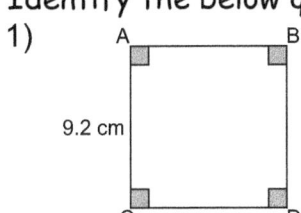

A) 169.28 cm² B) 84.64 cm²
C) 42.3 cm²   D) 85.44 cm²

2)

A) 21.6 mi² B) 38.5 mi²
C) 43.2 mi² D) 86.4 mi²

3)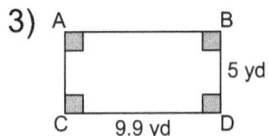

A) 41.8 yd² B) 49.5 yd²
C) 24.8 yd² D) 99 yd²

4)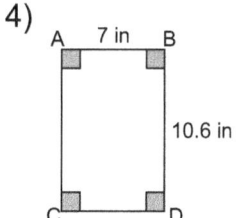

A) 82.9 in² B) 37.1 in²
C) 148.4 in² D) 74.2 in²

5)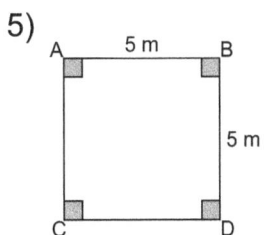

A) 25 m² B) 12.5 m²
C) 50 m² D) 21.1 m²

6)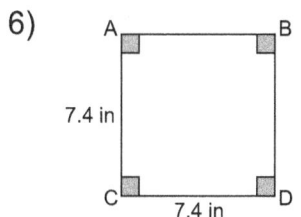

A) 57.16 in² B) 109.52 in²
C) 27.4 in²   D) 54.76 in²

Identify the below quadrilateral and find its area :

7)

A) 31.8 ft²   B) 25.4 ft²
C) 26.8 ft²   D) 15.9 ft²

8)

A) 4 yd²   B) 2 yd²
C) 8 yd²   D) 12.1 yd²

9)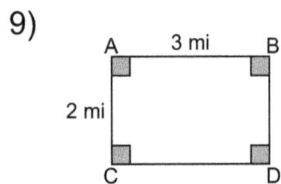

A) 3 mi²   B) 7.6 mi²
C) 6 mi²   D) 12 mi²

10)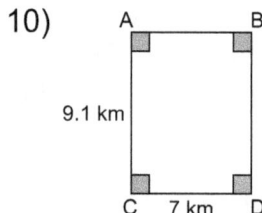

A) 72.3 km²   B) 76.4 km²
C) 31.9 km²   D) 63.7 km²

11)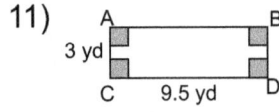

A) 14.3 yd²   B) 28.5 yd²
C) 27.5 yd²   D) 57 yd²

12)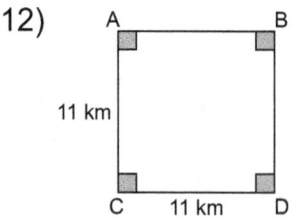

A) 126.5 km²   B) 242 km²
C) 121 km²   D) 60.5 km²

Identify the below quadrilateral and find its area:

13)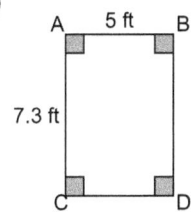

A) 36.5 ft²   B) 18.3 ft²
C) 73 ft²   D) 29.8 ft²

14)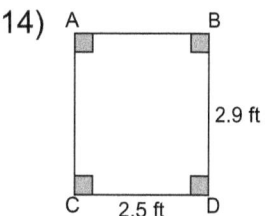

A) 3.6 ft²   B) 15.85 ft²
C) 7.25 ft²   D) 14.5 ft²

15)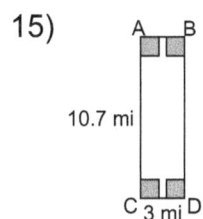

A) 32.1 mi²   B) 16.1 mi²
C) 25.4 mi²   D) 64.2 mi²

Identify the below quadrilateral and find its area :

1)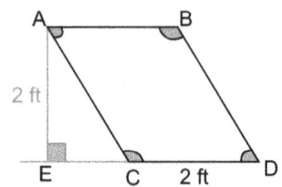

A) 2 ft²   B) 4 ft²
C) 11.7 ft²   D) 8 ft²

2)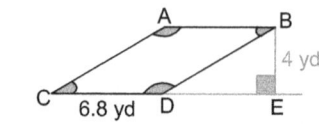

A) 27.2 yd²   B) 13.6 yd²
C) 54.4 yd²   D) 26 yd²

3)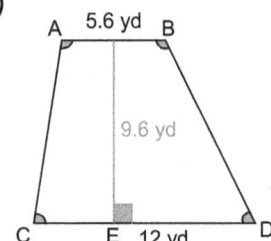

A) 84.48 yd²   B) 76.38 yd²
C) 79.18 yd²   D) 38.2 yd²

4)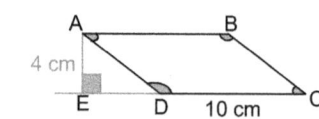

A) 20 cm²   B) 40 cm²
C) 80 cm²   D) 36.6 cm²

5)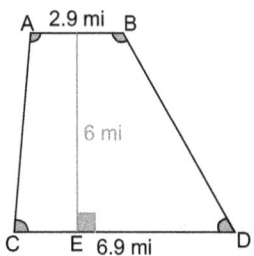

A) 42.6 mi²   B) 85.2 mi²
C) 29.4 mi²   D) 35.3 mi²

6)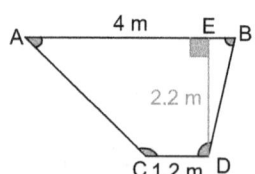

A) 11.52 m²   B) 11.44 m²
C) 2.9 m²   D) 5.72 m²

Identify the below quadrilateral and find its area :

7)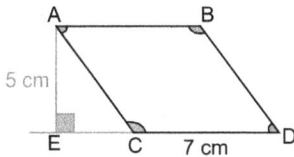

A) 17.5 cm² B) 70 cm²
C) 26.9 cm² D) 35 cm²

8)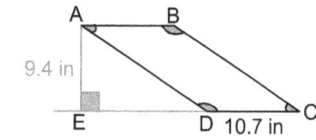

A) 90.68 in² B) 201.16 in²
C) 100.58 in² D) 95.18 in²

9)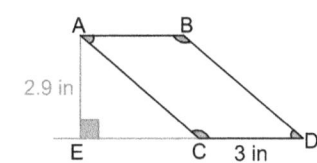

A) 4.4 in² B) 12.1 in²
C) 6.1 in² D) 8.7 in²

10)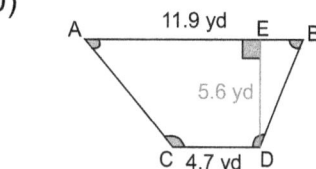

A) 95.76 yd² B) 92.96 yd²
C) 46.48 yd² D) 47.88 yd²

11)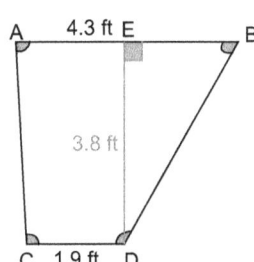

A) 17.78 ft² B) 5.9 ft²
C) 21.08 ft² D) 11.78 ft²

12)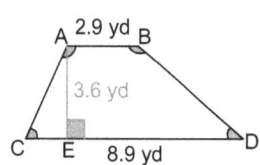

A) 21.24 yd² B) 12.3 yd²
C) 24.54 yd² D) 10.6 yd²

Parallelogram
Trapezium
#37

Identify the below quadrilateral and find its area :

13)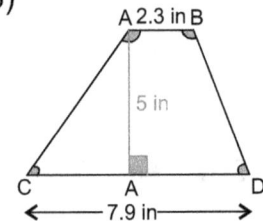

A) 29.6 in²   B) 12.8 in²
C) 25.5 in²   D) 51 in²

14)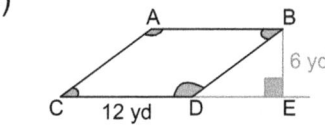

A) 78.2 yd²   B) 76 yd²
C) 72 yd²   D) 144 yd²

15)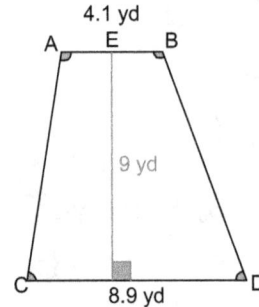

A) 117 yd²   B) 55.8 yd²
C) 29.3 yd²   D) 58.5 yd²

CIRCLE AREA CIRCUMFERENCE #38

1. Find the circumference of the circle given radius = 5.4 km and round the answer to the nearest tenth.

   (A) 33.9 km   (B) 145.6 km

   (C) 36.4 km   (D) 72.8 km

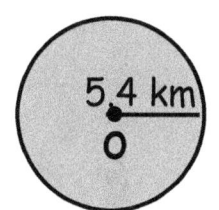

2. Find the circumference of the circle given radius = 6.9 in and round the answer to the nearest tenth.

   (A) 47.8 in   (B) 49.1 in

   (C) 49.7 in   (D) 43.3 in

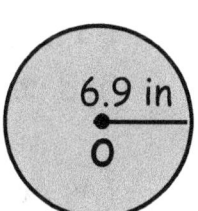

3. Find the circumference of the circle given radius = 5.3 mi and round the answer to the nearest tenth.

   (A) 20.5 mi   (B) 23.6 mi

   (C) 33.3 mi   (D) 66.6 mi

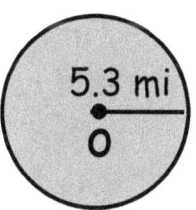

4. Find the circumference of the circle given radius = 3.7 yd and round the answer to the nearest tenth.

   (A) 23.2 yd   (B) 27 yd

   (C) 46.4 yd   (D) 32.7 yd

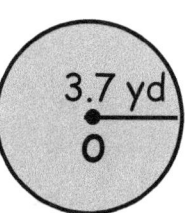

5. Find the area of the circle given radius = 3 cm and round the answer to the nearest tenth.

(A) 23.9 sq.cm   (B) 28.26 sq.cm

(C) 30.9 sq.cm   (D) 27.6 sq.cm

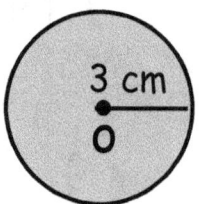

6. Find the circumference of the circle given radius = 2.3 ft and round the answer to the nearest tenth.

(A) 20.2 ft   (B) 14.4 ft

(C) 12.7 ft   (D) 15.2 ft

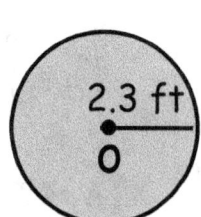

7. Find the circumference of the circle given radius = 5.8 in and round the answer to the nearest tenth.

(A) 36.4 in   (B) 59.7 in

(C) 18.2 in   (D) 52.7 in

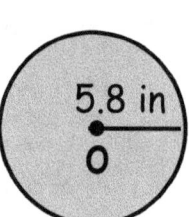

8. Find the circumference of the circle given radius = 2.9 ft and round the answer to the nearest tenth.

(A) 20.1 ft   (B) 18.2 ft

(C) 25.8 ft   (D) 27.7 ft

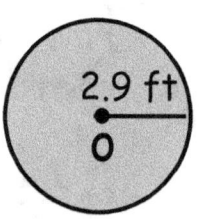

9. Find the circumference of the circle given radius = 10.3 yd and round the answer to the nearest tenth.

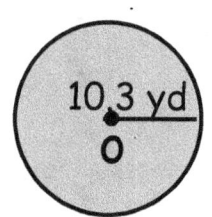

(A) 64.7 yd (B) 70.9 yd

(C) 75.9 yd (D) 67.8 yd

10. Find the circumference of the circle given radius = 9 cm and round the answer to the nearest tenth.

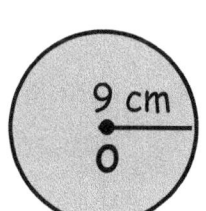

(A) 62.8 cm (B) 70.3 cm

(C) 56.5 cm (D) 65.3 cm

11. Find the circumference of the circle given radius = 6.6 m and round the answer to the nearest tenth.

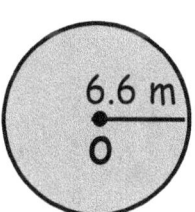

(A) 93.7 m (B) 88.7 m

(C) 41.5 m (D) 83 m

12. Find the circumference of the circle given radius = 7.7 yd and round the answer to the nearest tenth.

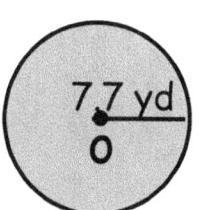

(A) 44 yd (B) 24.2 yd

(C) 48.4 yd (D) 27.3 yd

13. Find the circumference of the circle given radius = 6.1 in and round the answer to the nearest tenth.

    (A) 38.3 in (B) 38.9 in

    (C) 31.2 in (D) 15.6 in

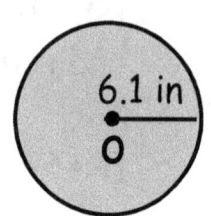

14. Find the circumference of the circle given radius = 2.4 in and round the answer to the nearest tenth.

    (A) 20.7 in (B) 17.6 in

    (C) 21.3 in (D) 15.1 in

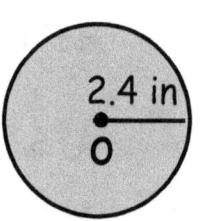

15. Find the circumference of the circle given radius = 8.5 m and round the answer to the nearest tenth.

    (A) 53.4 m (B) 25.1 m

    (C) 26.4 m (D) 31.4 m

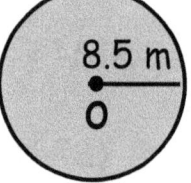

1. Find the volume of the sphere with a radius of 3 mi and round the answer to the nearest tenth.

   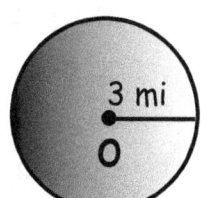

   (A) 43.3 mi³   (B) 96.1 mi³

   (C) 48.1 mi³   (D) 113.1 mi³

2. Find the volume of the sphere with a diameter of 25.2 ft and round the answer to the nearest tenth.

   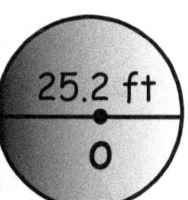

   (A) 30164.8 ft³   (B) 15082.4 ft³

   (C) 7541.2 ft³   (D) 8379.2 ft³

3. Find the volume of the sphere with a diameter of 9.2 ft and round the answer to the nearest tenth.

   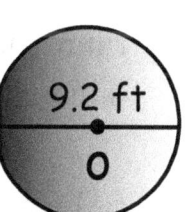

   (A) 265 ft³   (B) 338.4 ft³

   (C) 407.7 ft³   (D) 304.6 ft³

4. Find the volume of the sphere with a diameter of 11 in and round the answer to the nearest tenth.

   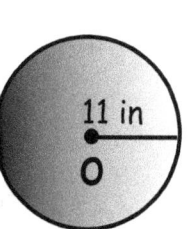

   (A) 696.9 in³   (B) 333.7 in³

   (C) 348.5 in³   (D) 292.7 in³

5. Find the volume of the sphere with a diameter of 37.2 m and round the answer to the nearest tenth.

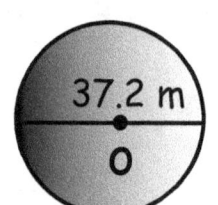

(A) 26954.3 m³

(B) 19474.4 mi³

(C) 15774.3 m³

(D) 22911.1 m³

6. Find the volume of the sphere with a diameter of 30 in and round the answer to the nearest tenth.

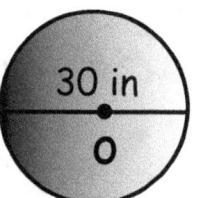

(A) 15550.9 in³

(B) 18350.1 in³

(C) 21836.6 in³

(D) 14137.2 in³

7. Find the volume of the sphere with a diameter of 23.2 cm and round the answer to the nearest tenth.

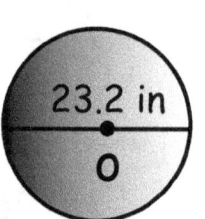

(A) 6538.3 cm³

(B) 5230.6 cm³

(C) 6119.8 cm³

(D) 5140.6 cm³

8. Find the volume of the sphere with a radius of 19.2 in and round the answer to the nearest tenth.

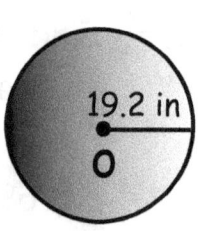

(A) 24014.7 in³

(B) 57155 in³

(C) 29647.8 in³

(D) 28577.5 in³

9. Find the volume of the sphere with a radius of 11 cm and round the answer to the nearest tenth.

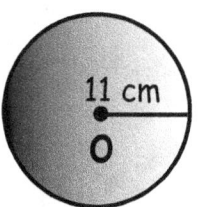

(A) 5575.3 cm^3

(B) 2805 cm^3

(C) 3261.6 cm^3

(D) 6523.1 cm^3

10. Find the volume of the sphere with a diameter of 23.4 cm and round the answer to the nearest tenth.

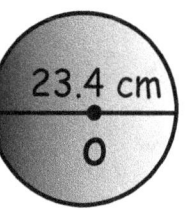

(A) 7379.7 cm^3

(B) 17563.6 cm^3

(C) 6708.8 cm^3

(D) 8781.8 cm^3

11. Find the volume of the sphere with a diameter of 16.4 cm and round the answer to the nearest tenth.

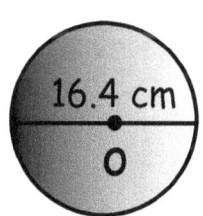

(A) 1385.8 m^3

(B) 1566 m^3

(C) 2309.6 m^3

(D) 1154.8 m^3

12. Find the volume of the sphere with a radius of 4.4 in and round the answer to the nearest tenth.

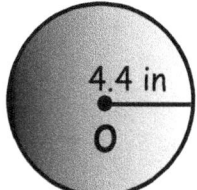

(A) 356.8 in^3

(B) 289 in^3

(C) 329.5 in^3

(D) 388.8 in^3

VOLUME SPHERE #39

13. Find the volume of the sphere with a radius of 6.6 cm and round the answer to the nearest tenth.

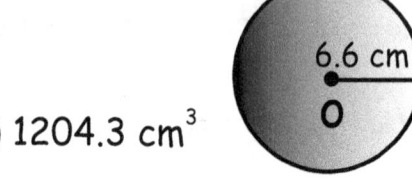

(A) 1820.8 cm^3

(B) 1204.3 cm^3

(C) 910.4 cm^3

(D) 1011.6 cm^3

14. Find the volume of the sphere with a diameter of 38.2 yd and round the answer to the nearest tenth.

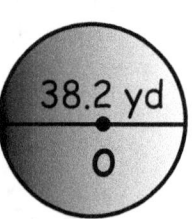

(A) 32105.6 yd^3

(B) 29186.9 yd^3

(C) 30648 yd^3

(D) 35637.2 yd^3

15. Find the volume of the sphere with a diameter of 5.4 ft and round the answer to the nearest tenth.

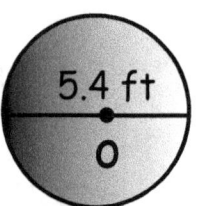

(A) 82.4 ft^3

(B) 70.1 ft^3

(C) 79.2 ft^3

(D) 91.9 ft^3

Pre-Algebra Vol 2

VOLUME RECTANGLE SQUARE #40

1. Find the volume of the rectangular prism measuring 2 cm and 6 cm along the base and 11 cm tall. Round the answer to the nearest tenth.

   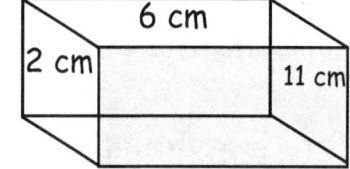

   (A) 79.9 cm³   (B) 105.6 cm³

   (C) 90.8 cm³   (D) 132 cm³

2. Find the volume of a square prism measuring 19 yd along each edge of the base and 11 yd tall. Round the answer to the nearest tenth.

   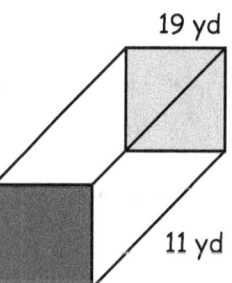

   (A) 3971 yd³   (B) 4447.5 yd³

   (C) 7827.6 yd³   (D) 3913.8 yd³

3. Find the volume of a rectangular prism measuring 11 in and 5 in along the base and 2 in tall. Round the answer to the nearest tenth.

   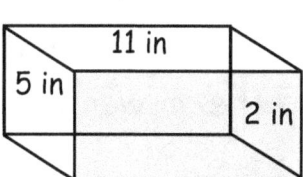

   (A) 62.7 in³   (B) 55 in³

   (C) 110 in³   (D) 131.6 mi³

4. Find the volume of a square prism measuring 7 mi along each edge of the base and 19 mi tall. Round the answer to the nearest tenth.

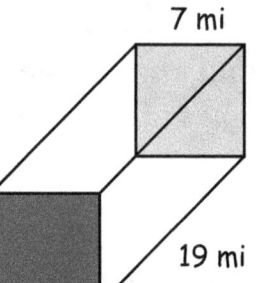

(A) 553.9 mi^3 (B) 465.5 mi^3

(C) 931 mi^3 (D) 1107.8 mi^3

5. Find the volume of a rectangular prism measuring 7 ft and 19 ft along the base and 4 ft tall. Round the answer to the nearest tenth.

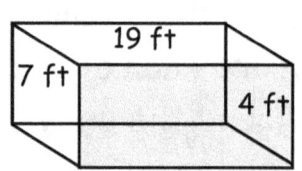

(A) 1064 ft^3 (B) 1170.4 ft^3

(C) 532 ft^3 (D) 2340.8 ft^3

6. Find the volume of a rectangular prism measuring 6 yd and 17 yd along the base and 7 yd tall. Round the answer to the nearest tenth.

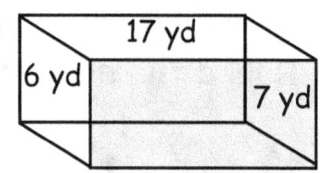

(A) 714 yd^3 (B) 1118.2 yd^3

(C) 1428 yd^3 (D) 1242.4 yd^3

7. Find the volume of a square prism measuring 13 ft along each edge of the base and 14 ft tall. Round the answer to the nearest tenth.

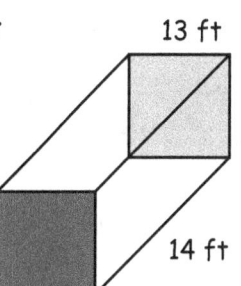

(A) 2366 ft^3

(B) 3236.6 ft^3

(C) 2697.2 ft^3

(D) 6473.2 ft^3

8. Find the volume of a rectangular prism measuring 4 mi and 7 mi along the base and 13 mi tall. Round the answer to the nearest tenth.

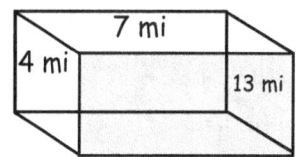

(A) 648 mi^3

(B) 324 mi^3

(C) 364 mi^3

(D) 719.3 mi^3

9. Find the volume of a square prism measuring 5 yd along each edge of the base and 4 yd tall. Round the answer to the nearest tenth.

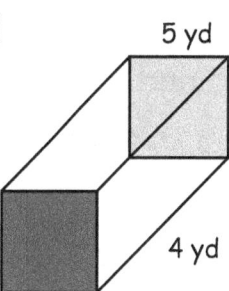

(A) 266.4 yd^3

(B) 200 yd^3

(C) 100 yd^3

(D) 222 yd^3

10. Find the volume of a square prism measuring 7 ft along each edge of the base and 8 ft tall. Round the answer to the nearest tenth.

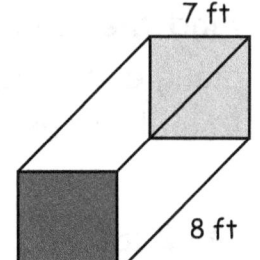

(A) 392 ft³   (B) 593.4 ft³

(C) 345 ft³   (D) 690 ft³

11. Find the volume of a square prism measuring 8 km along each edge of the base and 14 km tall. Round the answer to the nearest tenth.

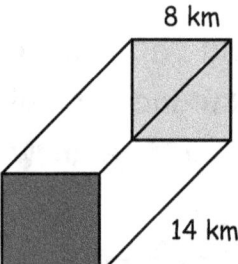

(A) 1015.1 km³   (B) 1066.2 km³

(C) 853 km³   (D) 896 km³

12. Find the volume of a rectangular prism measuring 18 yd and 15 yd along the base and 6 yd tall. Round the answer to the nearest tenth.

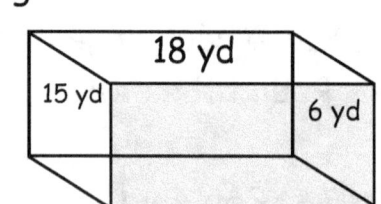

(A) 1409.4 yd³   (B) 1620 yd³

(C) 704.7 yd³   (D) 591.9 yd³

13. Find the volume of a rectangular prism measuring 5 m and 14 m along the base and 2 m tall. Round the answer to the nearest tenth.

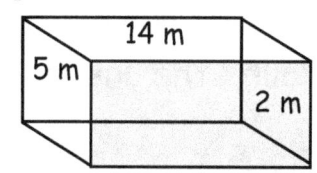

(A) 29.8 m³ (B) 59.5 m³

(C) 119 m³ (D) 140 m³

14. Find the volume of a rectangular prism measuring 19 cm and 8 cm along the base and 5 cm tall. Round the answer to the nearest tenth.

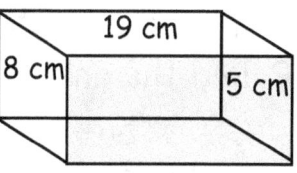

(A) 190 cm³ (B) 157.7 cm³

(C) 760 cm³ (D) 380 cm³

15. Find the volume of a rectangular prism measuring 13 cm and 20 cm along the base and 8 cm tall. Round the answer to the nearest tenth.

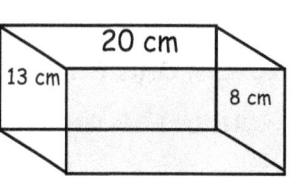

(A) 2080 cm³ (B) 3369.6 cm³

(C) 2931.6 cm³ (D) 4160 cm³

1. Find the volume of a cylinder with a diameter of 16 ft and a height of 4 ft.
   Round the answer to the nearest tenth.

   (A) 1608.5 ft³ (B) 3635.2 ft³

   (C) 3217 ft³   (D) 804.2 ft³

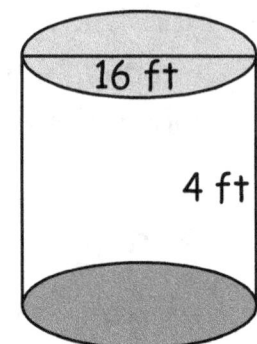

2. Find the volume of a cylinder with a diameter of 14 ft and a height of 16 ft.
   Round the answer to the nearest tenth.

   (A) 2615.8 ft³ (B) 2906.4 ft³

   (C) 3034.3 ft³ (D) 2463 ft³

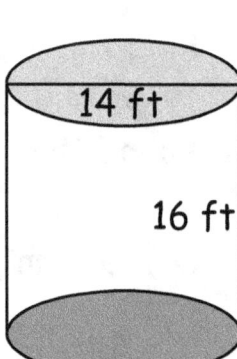

3. Find the volume of a cylinder with a diameter of 10 cm and a height of 10 cm.
   Round the answer to the nearest tenth.

   (A) 3141.6 cm³ (B) 2701.8 cm³

   (C) 1570.8 cm³ (D) 785.4 cm³

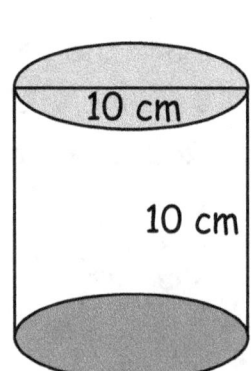

4. Find the volume of a cone with a diameter of 12 yd and a height of 14 yd.
Round the answer to the nearest tenth.

(A) 808 yd³

(B) 453.9 yd³

(C) 527.8 yd³

(D) 404 yd³

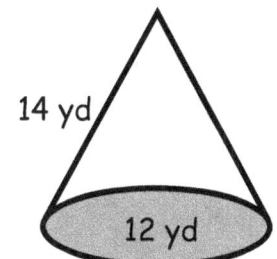

5. Find the volume of a cone with a diameter of 26 yd and a height of 26 yd.
Round the answer to the nearest tenth.

(A) 4601.4 yd³

(B) 7822.4 yd³

(C) 15644.8 yd³

(D) 3911.2 yd³

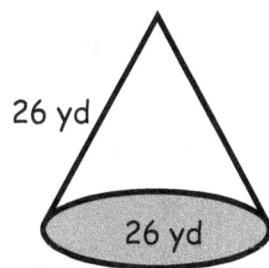

6. Find the volume of a cone with a diameter of 10 m and a height of 14 m.
Round the answer to the nearest tenth.

(A) 154 m³

(B) 366.5 m³

(C) 183.3 m³

(D) 77 m³

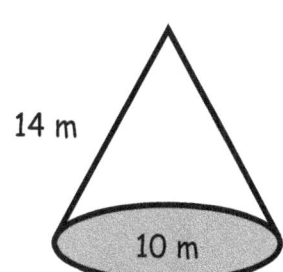

VOLUME CONE CYLINDER #41

7. Find the volume of a cylinder with a diameter of 16 ft and a height of 13 ft.
Round the answer to the nearest tenth.

   (A) 2529.1 ft^3 (B) 2613.8 ft^3

   (C) 2883.2 ft^3 (D) 2143.3 ft^3

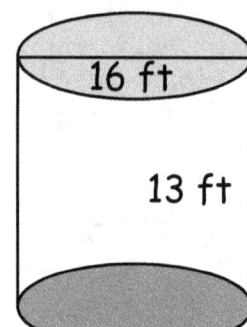

8. Find the volume of a cylinder with a diameter of 22 in and a height of 16 in.
Round the answer to the nearest tenth.

   (A) 14232.2 in^3 (B) 6082.1 in^3

   (C) 28464.4 in^3 (D) 7116.1 in^3

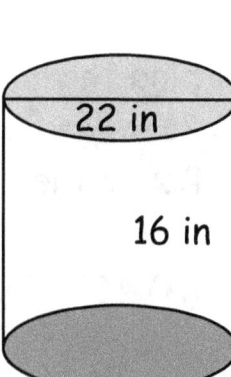

9. Find the volume of a cylinder with a radius of 14 m and a height of 16 m.
Round the answer to the nearest tenth.

   (A) 3514.3 m^3 (B) 8571.3 m^3

   (C) 9852 m^3 (D) 7028.5 m^3

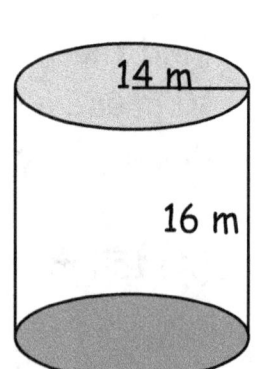

10. Find the volume of a cylinder with a radius of 10 cm and a height of 16 cm.
Round the answer to the nearest tenth.

(A) 3096.7 cm³

(B) 5026.5 cm³

(C) 2789.8 cm³

(D) 2513.3 cm³

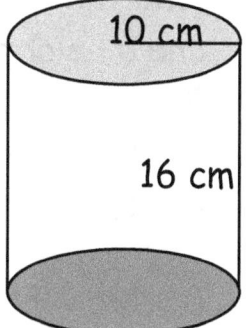

11. Find the volume of a cylinder with a radius of 9 ft and a height of 18 ft.
Round the answer to the nearest tenth.

(A) 3478.3 ft³

(B) 3664.4 ft³

(C) 4580.4 ft³

(D) 3078.1 ft³

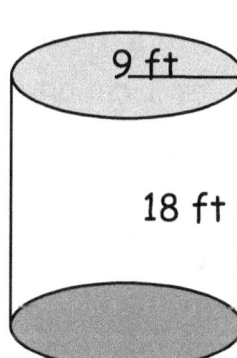

12. Find the volume of a cylinder with a diameter of 12 ft and a height of 19 ft.
Round the answer to the nearest tenth.

(A) 1891 ft³

(B) 945.5 ft³

(C) 784.8 ft³

(D) 2148.8 ft³

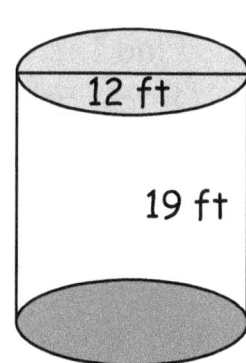

13. Find the volume of a cone with a radius of 3 ft and a height of 12 ft.
    Round the answer to the nearest tenth.

    (A) 113.1 ft³   (B) 110.2 ft³

    (C) 126.7 ft³   (D) 123.4 ft³

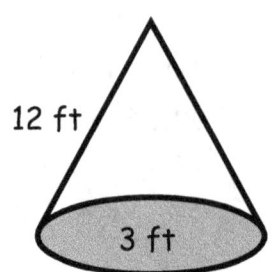

14. Find the volume of a cylinder with a radius of 1 cm and a height of 1 cm.
    Round the answer to the nearest tenth.

    (A) 2.5 cm³   (B) 5 cm³

    (C) 2.1 cm³   (D) 3.1 cm³

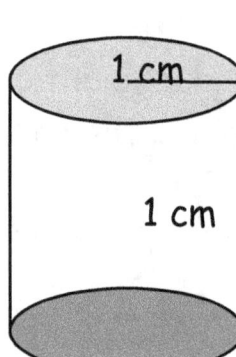

15. Find the volume of a cone with a diameter of 16 cm and a height of 20 cm.
    Round the answer to the nearest tenth.

    (A) 1340.4 cm³   (B) 1139.4 cm³

    (C) 649.5 cm³   (D) 569.7 cm³

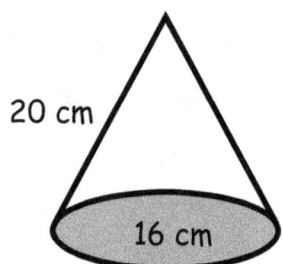

Pre-Algebra Vol 2

ADDITIVE INVERSE #42

Find the additive inverse of the below questions.

1. $\boxed{5}$

   (A) $\dfrac{1}{5}$ (B) $-\dfrac{1}{5}$ (C) -5 (D) 0

2. $\boxed{-66}$

   (A) 0 (B) $\dfrac{1}{66}$ (C) $-\dfrac{1}{66}$ (D) 66

3. $\boxed{-909}$

   (A) $\dfrac{1}{909}$ (B) 0 (C) 909 (D) $-\dfrac{1}{909}$

4. $\boxed{564}$

   (A) 0 (B) -564 (C) $\dfrac{1}{564}$ (D) $-\dfrac{1}{564}$

Find the additive inverse of the below questions.

5. 888

(A) -888 (B) $\frac{1}{888}$ (C) 0 (D) $-\frac{1}{888}$

6. -467

(A) 0 (B) $\frac{1}{467}$ (C) 467 (D) $-\frac{1}{467}$

7. 333

(A) $\frac{1}{333}$ (B) 0 (C) $-\frac{1}{333}$ (D) -333

8. $\frac{1}{6}$

(A) $-\frac{1}{6}$ (B) 0 (C) 6 (D) -6

Find the additive inverse of the below questions.

9. $\boxed{-\dfrac{8}{9}}$

(A) 0   (B) $\dfrac{8}{9}$ (C) $-\dfrac{9}{8}$ (D) $\dfrac{9}{8}$

10. $\boxed{\dfrac{13}{25}}$

(A) $\dfrac{13}{25}$ (B) $-\dfrac{13}{25}$ (C) $\dfrac{25}{13}$ (D) $-1\dfrac{12}{3}$

11. $\boxed{\dfrac{9}{29}}$

(A) $\dfrac{29}{9}$ (B) 0   (C) $-\dfrac{9}{29}$ (D) $-\dfrac{29}{9}$

12. $\boxed{-\dfrac{37}{893}}$

(A) 0   (B) $\dfrac{893}{37}$ (C) $-\dfrac{893}{37}$ (D) $\dfrac{37}{893}$

Find the additive inverse of the below questions.

13. $\boxed{\dfrac{41}{505}}$

(A) $-\dfrac{41}{505}$ (B) $-\dfrac{505}{41}$ (C) $\dfrac{505}{41}$ (D) 0

14. $\boxed{-\dfrac{2}{5}}$

(A) 0 (B) $\dfrac{2}{5}$ (C) $\dfrac{5}{2}$ (D) $-\dfrac{5}{2}$

15. $\boxed{\dfrac{5}{212}}$

(A) $-\dfrac{5}{212}$ (B) 0 (C) $\dfrac{212}{5}$ (D) $-\dfrac{212}{5}$

MULTIPLICATIVE INVERSE #43

Find the multiplicative inverse of the below questions.

1. $\boxed{21}$

   (A) −21   (B) $-\dfrac{1}{21}$ (C) 0   (D) $\dfrac{1}{21}$

2. $\boxed{-55}$

   (A) $-\dfrac{1}{55}$ (B) 0   (C) $\dfrac{1}{55}$ (D) 55

3. $\boxed{-301}$

   (A) 301   (B) $\dfrac{1}{301}$ (C) $-\dfrac{1}{301}$ (D) 0

4. $\boxed{601}$

   (A) $-\dfrac{1}{601}$ (B) $\dfrac{1}{601}$ (C) −601   (D) 0

MULTIPLICATIVE INVERSE #43

Find the multiplicative inverse of the below questions.

5. $\boxed{777}$

   (A) -777 (B) $-\dfrac{1}{777}$ (C) 0 (D) $\dfrac{1}{777}$

6. $\boxed{82}$

   (A) $-\dfrac{1}{82}$ (B) -82 (C) $\dfrac{1}{82}$ (D) 0

7. $\boxed{97}$

   (A) $\dfrac{1}{97}$ (B) $-\dfrac{1}{97}$ (C) -97 (D) 0

8. $\boxed{\dfrac{1}{7}}$

   (A) 7 (B) $-\dfrac{1}{7}$ (C) 0 (D) -7

Find the multiplicative inverse of the below questions.

9. $\boxed{-\dfrac{1}{8}}$

(A) 8   (B) 0   (C) $\dfrac{1}{8}$ (D) -8

10. $\boxed{\dfrac{1}{6}}$

(A) 0   (B) 6   (C) $-\dfrac{1}{6}$ (D) -6

11. $\boxed{-\dfrac{5}{7}}$

(A) -1.4   (B) 0   (C) $\dfrac{5}{7}$ (D) $-\dfrac{5}{7}$

12. $\boxed{99}$

(A) -99   (B) $-\dfrac{1}{99}$ (C) $\dfrac{1}{99}$ (D) 0

Find the multiplicative inverse of the below questions.

13. $\boxed{\dfrac{31}{69}}$

(A) 0   (B) $-\dfrac{31}{69}$ (C) –2.225806   (D) $2\dfrac{7}{31}$

14. $\boxed{-212}$

(A) 212   (B) $\dfrac{-1}{212}$ (C) 0   (D) 0.004717

15. $\boxed{565}$

(A) 0   (B) $-\dfrac{1}{565}$ (C) $\dfrac{1}{565}$ (D) –565

RATIONAL & IRRATIONAL NUMBERS #44

1. Which of the below is an irrational number?

   (A) $\sqrt{100}$ (B) $\sqrt{131}$ (C) $\sqrt{16}$ (D) $\sqrt{25}$

2. Which of the below is an irrational number?

   (A) $\sqrt{36}$ (B) $\sqrt{3}$ (C) $\sqrt{441}$ (D) $\sqrt{64}$

3. Which of the below is an irrational number?

   (A) $\sqrt{8}$ (B) $\sqrt{25}$ (C) $\sqrt{36}$ (D) $\sqrt{64}$

4. Which of the below is an irrational number?

   (A) $\sqrt{225}$ (B) $\sqrt{3600}$ (C) $\sqrt{81}$ (D) $\sqrt{43}$

5. Which of the below is an irrational number?

   (A) $\sqrt{77}$ (B) $\sqrt{4}$ (C) $\sqrt{9}$ (D) $\sqrt{121}$

6. Which of the below is an irrational number ?

   (A) $\sqrt{144}$ (B) $\sqrt{400}$ (C) $\sqrt{87}$ (D) $\sqrt{49}$

7. Which of the below is an irrational number ?

   (A) $\sqrt{2500}$ (B) $\sqrt{196}$ (C) $\sqrt{317}$ (D) $\sqrt{1089}$

8. Which of the below is an irrational number ?

   (A) $\sqrt{2500}$ (B) $\sqrt{963}$ (C) $\sqrt{169}$ (D) $\sqrt{11}$

9. Which of the below is an irrational number ?

   (A) $\sqrt{81}$ (B) $\sqrt{45}$ (C) $\sqrt{49}$ (D) $\sqrt{900}$

10. Which of the below is a rational number ?

    (A) $\sqrt{100}$ (B) $\sqrt{210}$ (C) $\sqrt{59}$ (D) $\sqrt{38}$

11. Which of the below is a rational number?

   (A) $\sqrt{36}$　　(B) $\sqrt{26}$　　(C) $\sqrt{46}$　　(D) $\sqrt{98}$

12. Which of the below is a rational number?

   (A) $\sqrt{101}$　　(B) $\sqrt{220}$　　(C) $\sqrt{144}$　　(D) $\sqrt{350}$

13. Which of the below is a rational number?

   (A) $\sqrt{444}$　　(B) $\sqrt{625}$　　(C) $\sqrt{560}$　　(D) $\sqrt{289}$

14. Which of the below is a rational number?

   (A) $\sqrt{345}$　　(B) $\sqrt{450}$　　(C) $\sqrt{92}$　　(D) $\sqrt{900}$

15. Which of the below is a rational number?

   (A) $\sqrt{29}$　　(B) $\sqrt{13}$　　(C) $\sqrt{8}$　　(D) $\sqrt{16}$

EXPONENTS #45

Pre-Algebra Vol 2

Simplify the below. Your answer should contain only positive exponents.

1. $4x \cdot 2x$

   (A) $8x^2$ (B) 3 (C) $6x^2$ (D) x^2

2. $2n^2 \cdot n^2$

   (A) $4n^2$ (B) $12n^2$ (C) $2n^3$ (D) $2n^4$

3. $2x^2 \cdot 2x$

   (A) $9x^3$ (B) $4x^3$ (C) $3x^4$ (D) $2x^2$

4. $3x^2 \cdot 2x^2$

   (A) $6x^4$ (B) $3x^4$ (C) $8x^3$ (D) $2x^3$

EXPONENTS #45

Simplify the below. Your answer should contain only positive exponents.

5. $3r \cdot 2r$

(A) $4r^2$ (B) $3r^3$ (C) $6r^2$ (D) r^2

6. $x^2 \cdot 3x^2$

(A) $3x^4$ (B) $6x^3$ (C) $8x^2$ (D) $4x^3$

7. $4x^2 \cdot 2x$

(A) $32x^3$ (B) $6x^2$ (C) $6x^3$ (D) $8x^3$

8. $2n^2 \cdot 4n^0$

(A) n^2 (B) $8n^2$ (C) $32n^3$ (D) n^3

EXPONENTS #45

Simplify the below. Your answer should contain only positive exponents.

9. $2k^0 \cdot k \cdot 4k$

   (A) $2k^4$ (B) $12k^4$ (C) $4k^4$ (D) $8k^2$

10. $2aa^0$

    (A) $12a^3$ (B) $2a$ (C) $4a^4$ (D) $4a^5$

11. $k \cdot 3k$

    (A) $6k^2$ (B) $3k$ (C) $3k^2$ (D) $4k^2$

12. $2n \cdot n$

    (A) $2n^2$ (B) $16n^4$ (C) $16n^3$ (D) $6n$

Simplify the below. Your answer should contain only positive exponents.

13. $n^2 \cdot 3n$

   (A) $6n^2$ (B) $3n^3$ (C) n^2 (D) $6n^3$

14. $4v^2 \cdot 3v \cdot 4v^2$

   (A) $3v^3$ (B) $48v^5$ (C) $4v^2$ (D) $3v^4$

15. $3x \cdot 3x$

   (A) $12x^3$ (B) $12x^5$ (C) $6x^3$ (D) $9x^2$

Find the slope of the straight line passing through the points given below.

1. A(18, −4), B(−16, −1)

    (A) $-\dfrac{34}{3}$ (B) $\dfrac{34}{3}$

    (C) $\dfrac{3}{34}$ (D) $-\dfrac{3}{34}$

2. A(−15, −1), B(−6, 6)

    (A) $-\dfrac{7}{9}$ (B) $\dfrac{7}{9}$

    (C) $-\dfrac{9}{7}$ (D) $\dfrac{9}{7}$

3. C(−15, −8), D(−9, 18)

    (A) $-\dfrac{3}{13}$ (B) $\dfrac{13}{3}$

    (C) $\dfrac{3}{13}$ (D) $-\dfrac{13}{3}$

Find the slope of the straight line passing through the points given below.

4. E(−6, 17), F(8, 17)

(A) $-\dfrac{3}{2}$ (B) Undefined

(C) $\dfrac{3}{2}$ (D) 0

5. G(−12, 11), H(−18, 11)

(A) $-\dfrac{2}{5}$ (B) $\dfrac{2}{5}$

(C) Undefined (D) 0

6. I(−5, −1), J(−18, −15)

(A) $-\dfrac{14}{13}$ (B) $\dfrac{13}{14}$

(C) $-\dfrac{13}{14}$ (D) $\dfrac{14}{13}$

Find the slope of the straight line passing through the points given below.

7. K(−5, −9), L(13, 8)

(A) $\dfrac{18}{17}$ (B) $-\dfrac{17}{18}$

(C) $-\dfrac{18}{17}$ (D) $\dfrac{17}{18}$

8. M(−4, 4), N(−7, 18)

(A) $-\dfrac{14}{3}$ (B) $\dfrac{14}{3}$

(C) $-\dfrac{3}{14}$ (D) $\dfrac{3}{14}$

9. O(12, 9), P(−3, 11)

(A) $\dfrac{2}{15}$ (B) $\dfrac{15}{2}$

(C) $-\dfrac{2}{15}$ (D) $-\dfrac{15}{2}$

Find the slope of the straight line passing through the points given below.

10. Q(–8, 2), R(–17, –12)

(A) $\dfrac{14}{9}$

(B) $-\dfrac{9}{14}$

(C) $\dfrac{9}{14}$

(D) $-\dfrac{14}{9}$

11. S(–9, –14), T(–2, 10)

(A) $-\dfrac{24}{7}$

(B) $-\dfrac{7}{24}$

(C) $\dfrac{24}{7}$

(D) $\dfrac{7}{24}$

12. U(–7, –1), V(16, –11)

(A) $-\dfrac{23}{10}$

(B) $\dfrac{10}{23}$

(C) $-\dfrac{10}{23}$

(D) $\dfrac{23}{10}$

Find the slope of the straight line passing through the points given below.

13. X(10, −17), Y(−9, −16)

(A) $\dfrac{1}{19}$ (B) 19

(C) $-\dfrac{1}{19}$ (D) −19

14. A(20, 19), B(−18, 6)

(A) $\dfrac{13}{38}$ (B) $-\dfrac{13}{38}$

(C) $\dfrac{38}{13}$ (D) $-\dfrac{38}{13}$

15. C(−2, 12), D(−19, 2)

(A) $-\dfrac{17}{10}$ (B) $\dfrac{10}{17}$

(C) $\dfrac{17}{10}$ (D) $-\dfrac{10}{17}$

Pre-Algebra Vol 2

SLOPE INTERCEPT FORM #47

Write the slope-intercept form of the equation of each straight line given the slope and y-intercept.

1. Slope = $-\dfrac{4}{3}$, y-intercept = -1

(A) $y = \dfrac{4}{3}x - 1$

(B) $y = -x + \dfrac{4}{3}$

(C) $y = -x - \dfrac{4}{3}$

(D) $y = -\dfrac{4}{3}x - 1$

2. Slope = -1, y-intercept = 1

(A) $y = -x + 1$

(B) $y = x + 1$

(C) $y = -4x + 1$

(D) $y = 4x + 1$

3. Slope = $\dfrac{3}{2}$, y-intercept = 3

(A) $y = -x + 3$

(B) $y = 3x + 3$

(C) $y = -\dfrac{3}{2}x + 3$

(D) $y = \dfrac{3}{2}x + 3$

SLOPE INTERCEPT FORM #47

Write the slope-intercept form of the equation of each straight line given the slope and y-intercept.

4. Slope = $-\dfrac{9}{4}$, y-intercept = 5

(A) $y = -\dfrac{9}{4}x + 5$ (B) $y = -\dfrac{1}{4}x + 5$

(C) $y = \dfrac{1}{2}x + 5$ (D) $y = \dfrac{1}{4}x + 5$

5. Slope = $-\dfrac{3}{2}$, y-intercept = -3

(A) $y = -3x + \dfrac{5}{2}$ (B) $y = \dfrac{5}{2}x - 3$

(C) $y = -\dfrac{5}{2}x - 3$ (D) $y = -\dfrac{3}{2}x - 3$

6. Slope = $-\dfrac{5}{2}$, y-intercept = 5

(A) $y = -2x + 5$ (B) $y = x + 5$

(C) $y = -\dfrac{5}{2}x + 5$ (D) $y = 2x + 5$

SLOPE INTERCEPT FORM #47

Write the slope-intercept form of the equation of each straight line given the slope and y-intercept.

7. Slope = $\dfrac{3}{2}$, y-intercept = 1

(A) $y = \dfrac{1}{2}x + 1$ (B) $y = \dfrac{3}{2}x + 1$

(C) $y = -\dfrac{1}{2}x + 1$ (D) $y = -\dfrac{3}{2}x + 1$

8. Slope = 6, y-intercept = 2

(A) $y = 4x + 2$ (B) $y = -6x + 2$

(C) $y = 6x + 2$ (D) $y = -4x + 2$

9. Slope = $\dfrac{1}{4}$, y-intercept = 0

(A) $y = \dfrac{1}{4}x$ (B) $y = \dfrac{4}{3}x$

(C) $y = \dfrac{4}{3}x - \dfrac{1}{3}$ (D) $y = -\dfrac{1}{4}x$

Write the slope-intercept form of the equation of each straight line given the slope and y-intercept.

10. Slope = $-\dfrac{3}{2}$, y-intercept = -2

(A) $y = -\dfrac{5}{2}x - 2$

(B) $y = -\dfrac{3}{2}x - 2$

(C) $y = \dfrac{3}{2}x - 2$

(D) $y = \dfrac{5}{2}x - 2$

11. Slope = $-\dfrac{1}{4}$, y-intercept = 5

(A) $y = -\dfrac{1}{4}x - \dfrac{1}{4}$

(B) $y = -\dfrac{1}{4}x + 5$

(C) $y = \dfrac{1}{4}x - \dfrac{1}{4}$

(D) $y = 5x - \dfrac{1}{4}$

12. Slope = $\dfrac{7}{5}$, y-intercept = -4

(A) $y = \dfrac{7}{5}x - 4$

(B) $y = 4x + \dfrac{7}{5}$

(C) $y = -4x + \dfrac{7}{5}$

(D) $y = -\dfrac{7}{5}x - 4$

SLOPE INTERCEPT FORM #47

Write the slope-intercept form of the equation of each straight line given the slope and y-intercept.

13. Slope = –7, y-intercept = 2

(A) y = 2x + 5

(B) y = –5x + 2

(C) y = –7x + 2

(D) y = 5x + 2

14. Slope = 5, y-intercept = 1

(A) y = –5x + 1

(B) y = x + 5

(C) y = 5x + 1

(D) y = –x + 5

15. Slope = $\frac{1}{5}$, y-intercept = 3

(A) y = –x + 3

(B) y = $\frac{4}{5}$x + 3

(C) y = $\frac{1}{5}$x + 3

(D) y = –$\frac{1}{5}$x + 3

Find the slope of the straight line for each of the questions given below :

1)
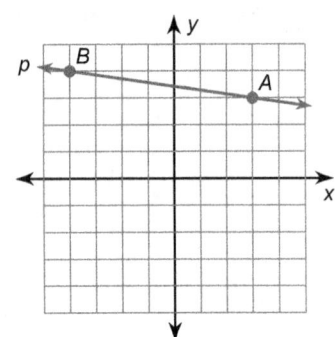

A) $\dfrac{1}{7}$ B) -7

C) 7 D) $-\dfrac{1}{7}$

2)
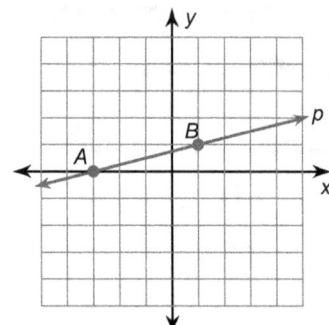

A) $\dfrac{1}{4}$ B) $-\dfrac{1}{4}$

C) 4 D) -4

3)
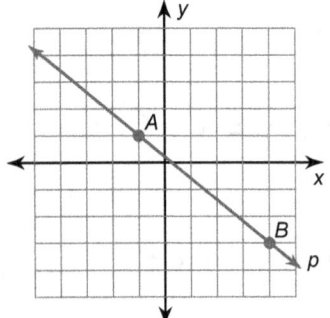

A) $\dfrac{5}{4}$ B) $-\dfrac{5}{4}$

C) $-\dfrac{4}{5}$ D) $\dfrac{4}{5}$

4)
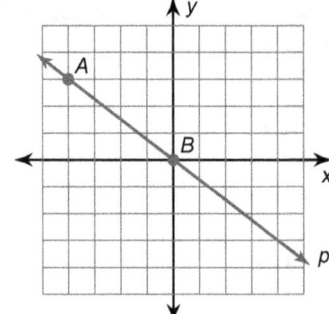

A) $-\dfrac{3}{4}$ B) $\dfrac{4}{3}$

C) $-\dfrac{4}{3}$ D) $\dfrac{3}{4}$

Find the slope of the straight line for each of the questions given below :

5)
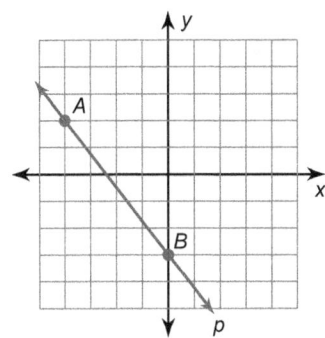

A) $\dfrac{5}{4}$ B) $-\dfrac{4}{5}$

C) $-\dfrac{5}{4}$ D) $\dfrac{4}{5}$

6)
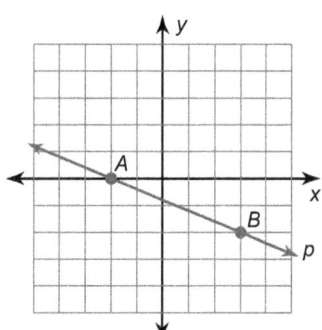

A) $-\dfrac{5}{2}$ B) $\dfrac{2}{5}$

C) $\dfrac{5}{2}$ D) $-\dfrac{2}{5}$

7)
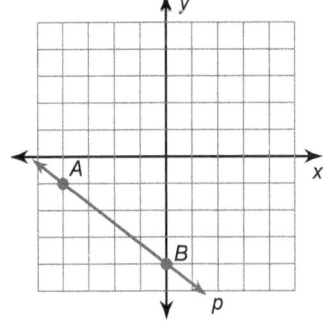

A) $-\dfrac{3}{4}$ B) $-\dfrac{4}{3}$

C) $\dfrac{3}{4}$ D) $\dfrac{4}{3}$

8)
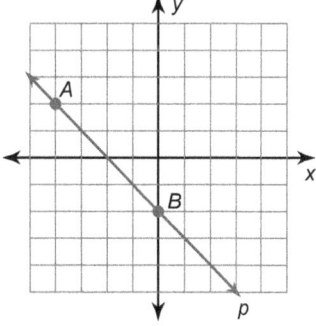

A) 1 B) $-\dfrac{3}{4}$

C) $\dfrac{3}{4}$ D) -1

Find the slope of the straight line for each of the questions given below :

9)
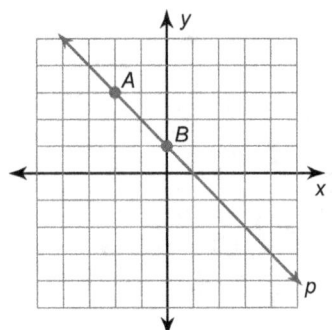

A) $-\dfrac{3}{4}$ B) $\dfrac{3}{4}$

C) 1   D) -1

10)
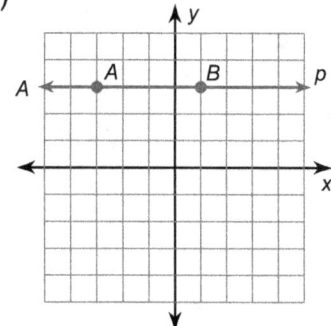

A) 2   B) -2

C) 0   D) Undefined

11)
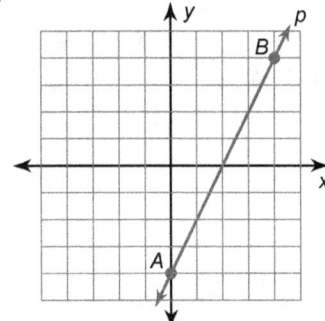

A) 2   B) -2

C) $\dfrac{1}{2}$ D) $-\dfrac{1}{2}$

12)
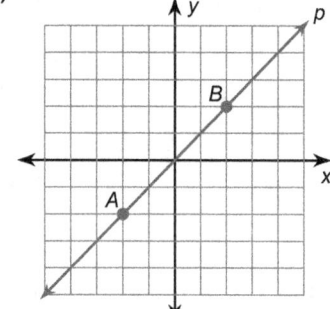

A) -1 B) 5

C) 1   D) -5

Find the slope of the straight line for each of the questions given below :

13)
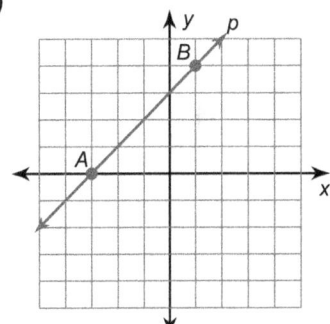

A) $-\dfrac{5}{3}$ B) 1

C) $\dfrac{5}{3}$ D) -1

14)
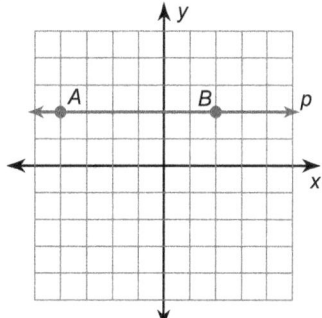

A) $-\dfrac{3}{4}$ B) Undefined

C) 0   D) $\dfrac{3}{4}$

15)
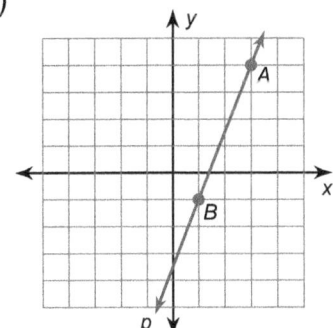

A) $\dfrac{2}{5}$ B) $\dfrac{5}{2}$

C) $-\dfrac{2}{5}$ D) $-\dfrac{5}{2}$

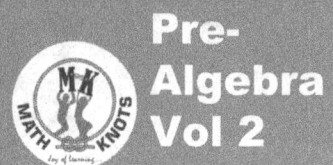

Find slope #49

Find the slope of the straight line for each of the questions given below :

1) $2x + y = -4$

   A) -2 B) $-\frac{1}{2}$ C) 2 D) $\frac{1}{2}$

2) $x + 4y = -20$

   A) 4 B) -4 C) $\frac{1}{4}$ D) $-\frac{1}{4}$

3) $2x + 3y = -15$

   A) $-\frac{2}{3}$ B) $-\frac{3}{2}$ C) $\frac{3}{2}$ D) $\frac{2}{3}$

4) $x + 5y = -25$

   A) $-\frac{1}{5}$ B) -5 C) $\frac{1}{5}$ D) 5

5) $5x - 2y = 0$

   A) $\frac{2}{5}$ B) $\frac{5}{2}$ C) $-\frac{5}{2}$ D) $-\frac{2}{5}$

Find slope #49

Find the slope of the straight line for each of the questions given below :

6) $x - y = -2$

   A) 1 B) -1 C) $\frac{3}{2}$ D) $-\frac{3}{2}$

7) $y = 0$

   A) 2 B) Undefined C) 0 D) -2

8) $7x + 3y = -15$

   A) $-\frac{7}{3}$ B) $\frac{3}{7}$ C) $-\frac{3}{7}$ D) $\frac{7}{3}$

9) $x = -1$

   A) 2 B) 0 C) Undefined D) -2

10) $7x - 4y = 16$

   A) $\frac{4}{7}$ B) $-\frac{4}{7}$ C) $\frac{7}{4}$ D) $-\frac{7}{4}$

Find the slope of the straight line for each of the questions given below :

11) $y = -2$

A) Undefined B) $-\dfrac{3}{5}$ C) 0 D) $\dfrac{3}{5}$

12) $x + y = -4$

A) $-\dfrac{4}{5}$ B) $\dfrac{4}{5}$ C) -1 D) 1

13) $4x - y = 4$

A) $-\dfrac{1}{4}$ B) -4 C) $\dfrac{1}{4}$ D) 4

14) $2x + 5y = 5$

A) $\dfrac{2}{5}$ B) $\dfrac{5}{2}$ C) $-\dfrac{2}{5}$ D) $-\dfrac{5}{2}$

15) $3x - y = 0$

A) -3 B) 3 C) $-\dfrac{1}{3}$ D) $\dfrac{1}{3}$

Parallel line slope #50

Find the slope of the straight line which is parallel to given straight line as below:

1) $2x - y = 2$

   A) 2 B) $-\dfrac{1}{2}$ C) $\dfrac{1}{2}$ D) -2

2) $4x - 3y = -3$

   A) $\dfrac{3}{4}$ B) $-\dfrac{3}{4}$ C) $-\dfrac{4}{3}$ D) $\dfrac{4}{3}$

3) $x = 2$

   A) $\dfrac{1}{4}$ B) $-\dfrac{1}{4}$ C) Undefined D) 0

4) $2x - 3y = -12$

   A) $-\dfrac{3}{2}$ B) $\dfrac{3}{2}$ C) $\dfrac{2}{3}$ D) $-\dfrac{2}{3}$

5) $10x - 3y = 15$

   A) $-\dfrac{10}{3}$ B) $\dfrac{3}{10}$ C) $-\dfrac{3}{10}$ D) $\dfrac{10}{3}$

Pre-Algebra Vol 2

Parallel line slope #50

Find the slope of the straight line which is parallel to given straight line as below:

6) $2x + y = 1$

   A) 2 B) $-\frac{1}{2}$ C) -2 D) $\frac{1}{2}$

7) $2x - y = 1$

   A) 2 B) $-\frac{1}{2}$ C) $\frac{1}{2}$ D) -2

8) $7x - 4y = -20$

   A) $-\frac{4}{7}$ B) $\frac{7}{4}$ C) $-\frac{7}{4}$ D) $\frac{4}{7}$

9) $8x - 5y = 20$

   A) $\frac{5}{8}$ B) $-\frac{8}{5}$ C) $\frac{8}{5}$ D) $-\frac{5}{8}$

10) $x + y = 3$

    A) 5 B) -5 C) 1 D) -1

11) $3x + 2y = 2$

    A) $\frac{3}{2}$ B) $-\frac{3}{2}$ C) $\frac{2}{3}$ D) $-\frac{2}{3}$

Parallel line slope #50

Find the slope of the straight line which is parallel to given straight line as below:

12) $5x + 4y = -16$

A) $\dfrac{4}{5}$ B) $\dfrac{5}{4}$ C) $-\dfrac{4}{5}$ D) $-\dfrac{5}{4}$

13) $3x + y = 1$

A) $-\dfrac{1}{3}$ B) -3 C) $\dfrac{1}{3}$ D) 3

14) $x + 2y = 2$

A) -2 B) $-\dfrac{1}{2}$ C) $\dfrac{1}{2}$ D) 2

15) $x + y = -5$

A) $\dfrac{3}{2}$ B) -1 C) $-\dfrac{3}{2}$ D) 1

Pre-Algebra Vol 2

Perpendicular slope #51

Find the slope of the straight line which is perpendicular to given straight line as below:

1) $y = \frac{7}{4}x - 4$

   A) $-\frac{4}{7}$ B) $\frac{7}{4}$ C) $\frac{4}{7}$ D) $-\frac{7}{4}$

2) $y = \frac{7}{4}x + 5$

   A) $-\frac{4}{7}$ B) $-\frac{7}{4}$ C) $\frac{7}{4}$ D) $\frac{4}{7}$

3) $y = -\frac{2}{3}x - 2$

   A) $\frac{3}{2}$ B) $-\frac{2}{3}$ C) $-\frac{3}{2}$ D) $\frac{2}{3}$

4) $y = -\frac{3}{5}x + 1$

   A) $-\frac{3}{5}$ B) $\frac{3}{5}$ C) $-\frac{5}{3}$ D) $\frac{5}{3}$

5) $y = 4x - 5$

   A) $\frac{1}{4}$ B) $-\frac{1}{4}$ C) -4 D) 4

Perpendicular slope #51

Find the slope of the straight line which is perpendicular to given straight line as below:

6) $y = \frac{9}{4}x + 4$

  A) $-\frac{9}{4}$ B) $\frac{9}{4}$ C) $\frac{4}{9}$ D) $-\frac{4}{9}$

7) $y = \frac{3}{4}x - 1$

  A) $\frac{3}{4}$ B) $-\frac{4}{3}$ C) $-\frac{3}{4}$ D) $\frac{4}{3}$

8) $y = -2x + 5$

  A) $\frac{1}{2}$ B) 2 C) $-\frac{1}{2}$ D) -2

9) $y = -4$

  A) $\frac{2}{5}$ B) Undefined   C) 0 D) $-\frac{2}{5}$

10) $y = -\frac{3}{5}x - 2$

  A) $\frac{3}{5}$ B) $-\frac{3}{5}$ C) $\frac{5}{3}$ D) $-\frac{5}{3}$

Find the slope of the straight line which is perpendicular to given straight line as below :

11) $y = \dfrac{5}{2}x$

A) $-\dfrac{2}{5}$ B) $-\dfrac{5}{2}$ C) $\dfrac{5}{2}$ D) $\dfrac{2}{5}$

12) $x = 5$

A) Undefined   B) $-\dfrac{1}{2}$ C) 0   D) $\dfrac{1}{2}$

13) $y = -5x + 5$

A) $-\dfrac{1}{5}$ B) -5 C) $\dfrac{1}{5}$ D) 5

14) $y = -3x - 2$

A) $\dfrac{1}{3}$ B) -3 C) 3   D) $-\dfrac{1}{3}$

15) $y = 4$

A) 2   B) 0   C) Undefined   D) -2

Find the measure of the ∠x for the below questions:

1)
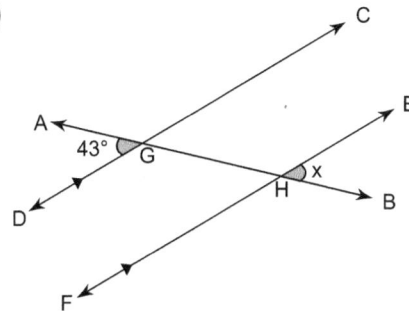

A) 136°   B) 46°
C) 134°   D) 43°

2)
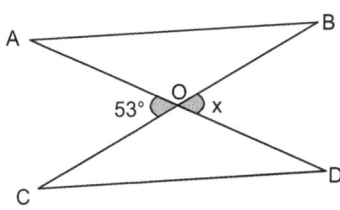

A) 47°   B) 27°
C) 143°   D) 53°

3)
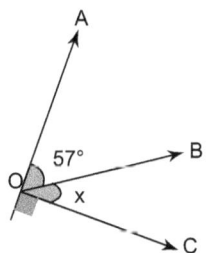

A) 137°   B) 33°
C) 147°   D) 46°

4)
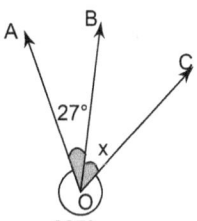

A) 36°   B) 154°
C) 146°   D) 138°

5)
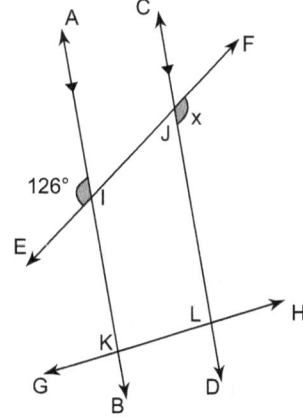

A) 130°   B) 54°
C) 154°   D) 126°

6)

A) 30°   B) 150°
C) 126°   D) 70°

Pre-Algebra Vol 2

Angle measure #52

Find the measure of the ∠x for the below questions:

7)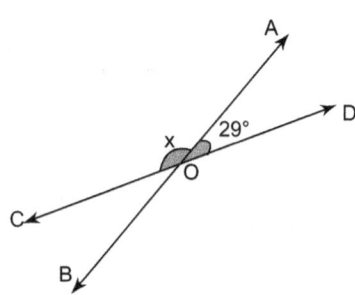

A) 151°   B) 161°
C) 51°   D) 141°

8)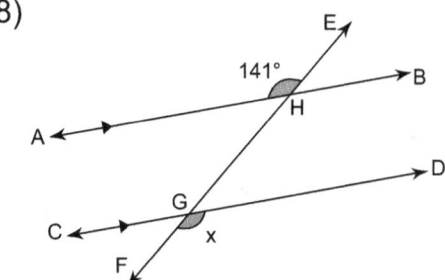

A) 29°   B) 154°
C) 39°   D) 141°

9)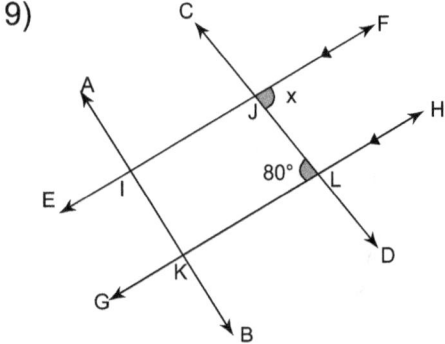

A) 80°   B) 110°
C) 10°   D) 120°

10)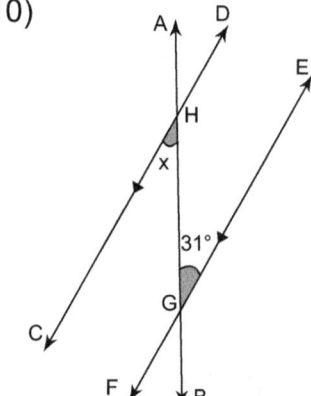

A) 149°   B) 31°
C) 49°   D) 59°

11)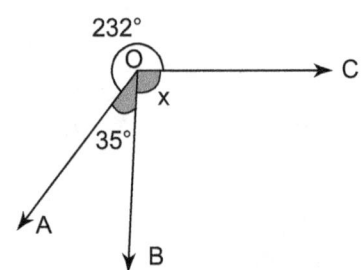

A) 93°   B) 147°
C) 83°   D) 7°

12)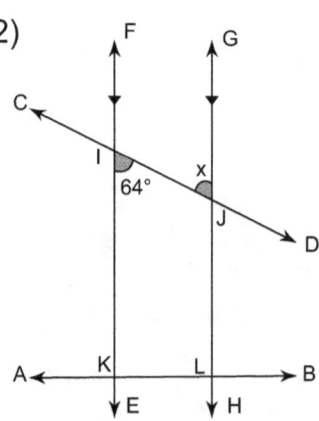

A) 126°   B) 64°
C) 138°   D) 26°

Find the measure of the ∠x for the below questions:

13)

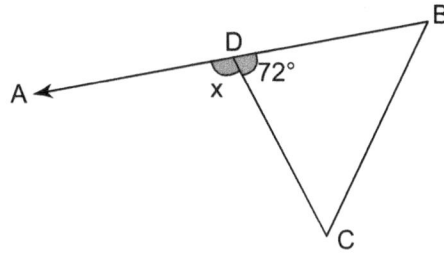

A) 48° B) 78°
C) 108° D) 42°

14)

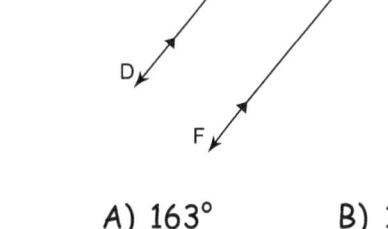

A) 163° B) 153°
C) 23° D) 47°

15)

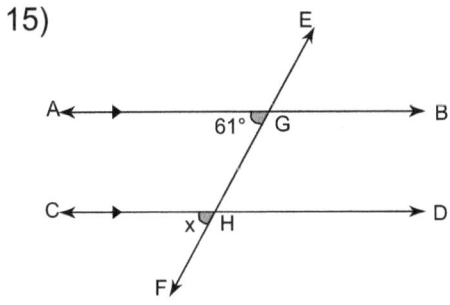

A) 61° B) 119°
C) 129° D) 149°

Pre-Algebra Vol 2

Angle measure #53

Find the measure of the ∠x for the below questions:

1)

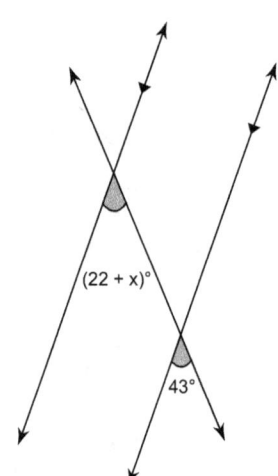

A) 21°   B) 20°
C) 25°   D) 16°

2)

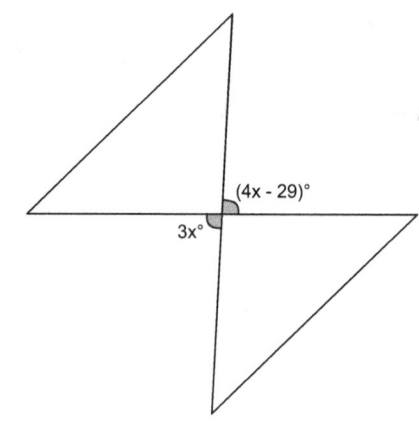

A) 19°   B) 18°
C) 29°   D) 21°

3)

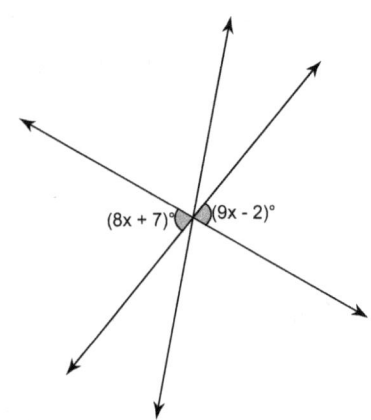

A) 9°   B) 11°
C) 7°   D) 13°

4)

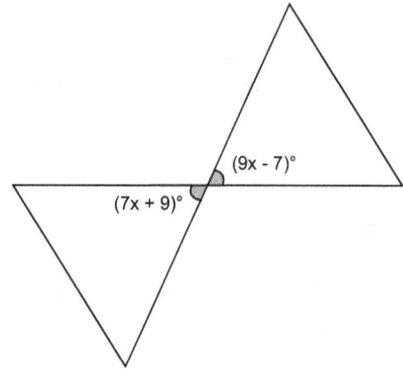

A) 15°   B) 13°
C) 18°   D) 8°

Find the measure of the x for the below questions:

5)

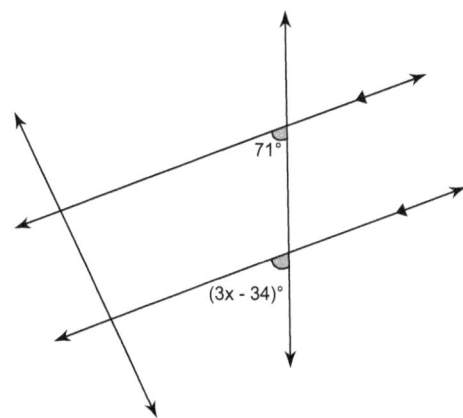

A) 35°   B) 39°
C) 25°   D) 28°

6)

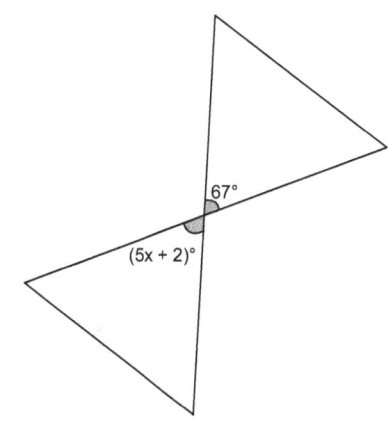

A) 11°   B) 13°
C) 17°   D) 6°

7)

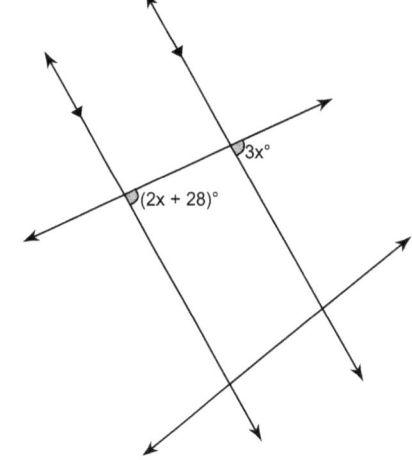

A) 28°   B) 24°
C) 33°   D) 26°

8)

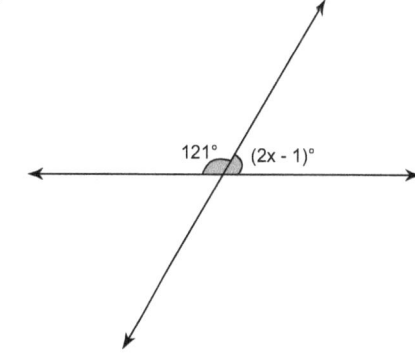

A) 30°   B) 33°
C) 28°   D) 29°

Find the measure of the ∠x for the below questions:

9)

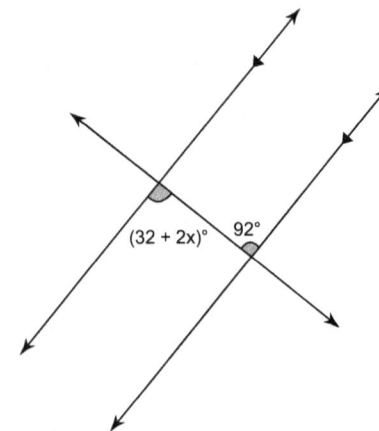

A) 28°   B) 25°
C) 32°   D) 30°

10)

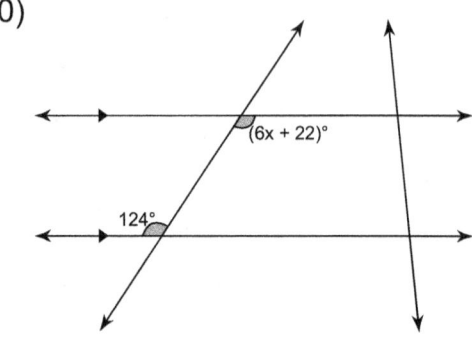

A) 15°   B) 13°
C) 17°   D) 21°

11)

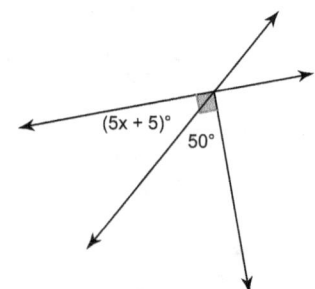

A) 3°   B) 5°
C) 7°   D) 9°

12)

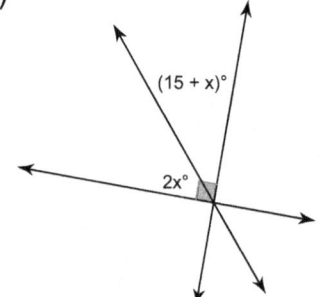

A) 19°   B) 21°
C) 28°   D) 25°

Find the measure of the ∠x for the below questions:

13)

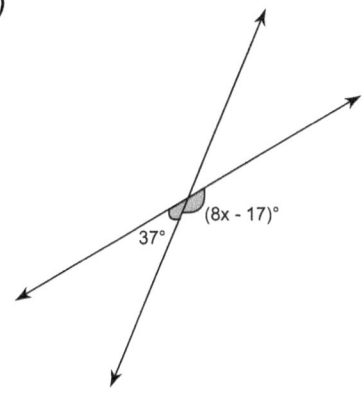

A) 20°   B) 18°
C) 15°   D) 4°

14)

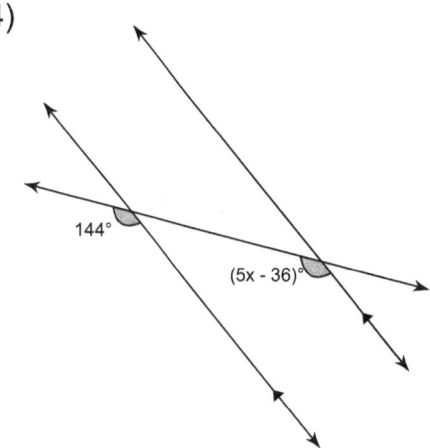

A) 39°   B) 40°
C) 36°   D) 33°

15)

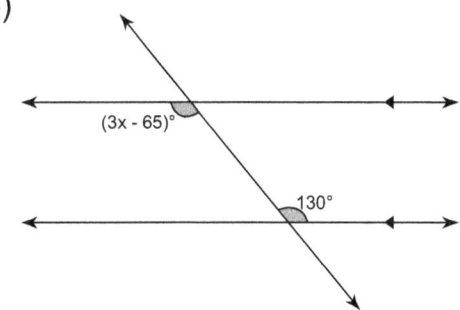

A) 66°   B) 59°
C) 55°   D) 65°

Describe the transformation and find the transformation rule:

1)
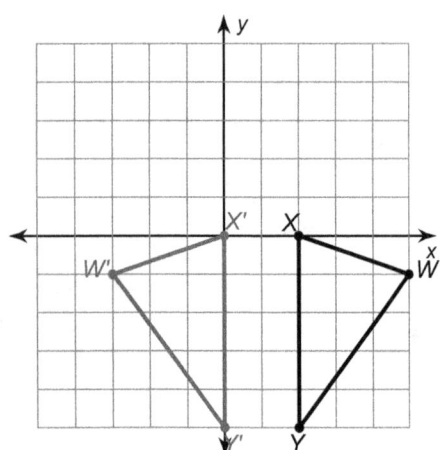

A) Reflection across y = -2
B) Reflection across x = -1
C) Reflection across x = 1
D) Reflection across the x-axis

2)
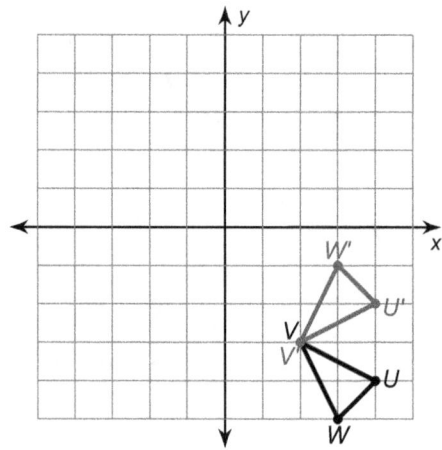

A) Reflection across y = -3
B) Reflection across y = 3
C) Reflection across x = -3
D) Reflection across x = 3

3)
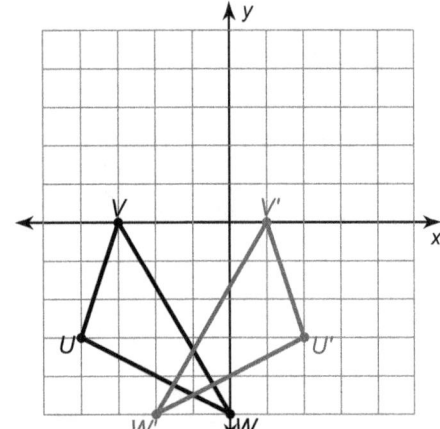

A) Reflection across the y-axis
B) Reflection across y = 1
C) Reflection across x = -1
D) Reflection across x = 2

4)
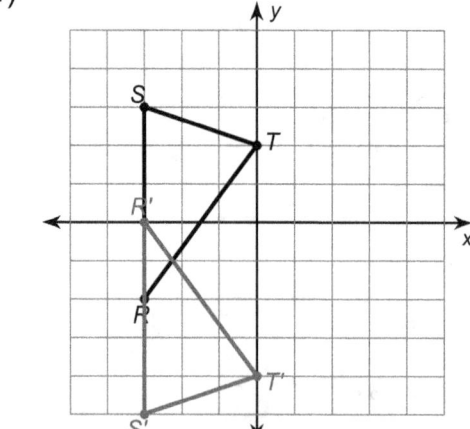

A) Reflection across x = 2
B) Reflection across the x-axis
C) Reflection across x = 1
D) Reflection across y = -1

Describe the transformation and find the transformation rule :

5)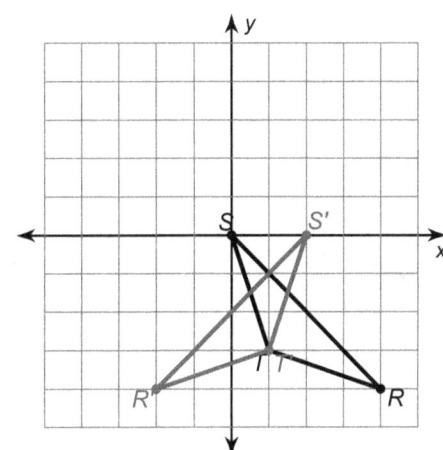

A) Reflection across y = 2
B) Reflection across the x-axis
C) Reflection across y = 1
D) Reflection across x = 1

6)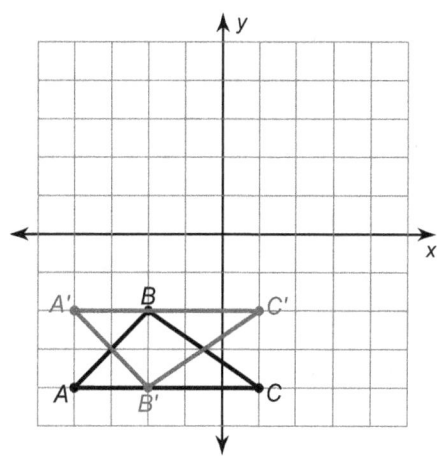

A) Reflection across y = –3
B) Reflection across x = 1
C) Reflection across = 3
D) Reflection across the x-axis

7)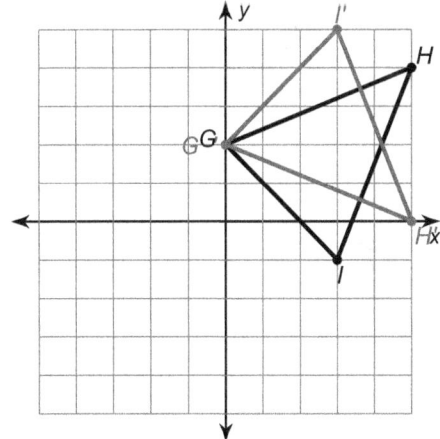

A) Reflection across the y-axis
B) Reflection across x = - 1
C) Reflection across x = - 2
D) Reflection across y = 2

8)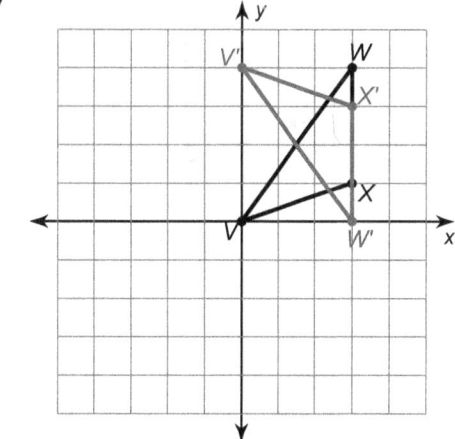

A) Reflection across y = 2
B) Reflection across the x-axis
C) Reflection across x = - 2
D) Reflection across y = - 1

Describe the transformation and find the transformation rule :

9)
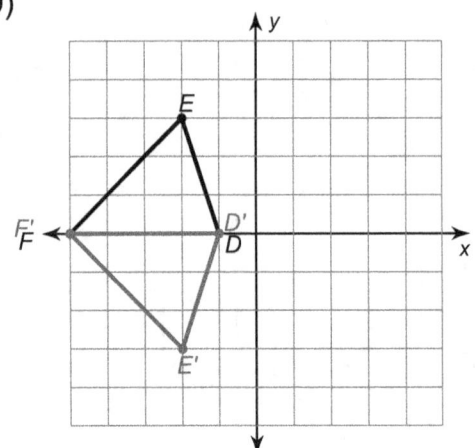

A) Reflection across x = 1
B) Reflection across y = - 1
C) Reflection across y = 5
D) Reflection across the x-axis

10)
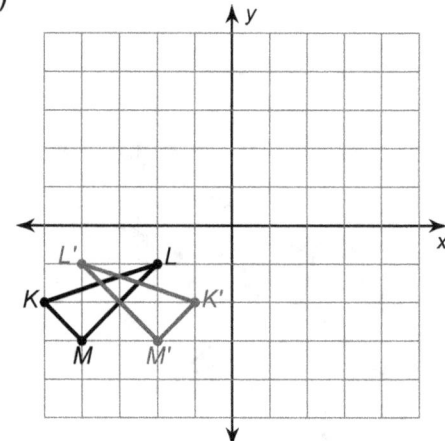

A) Reflection across y = 1
B) Reflection across y = 3
C) Reflection across x = -1
D) Reflection across x = -3

11)
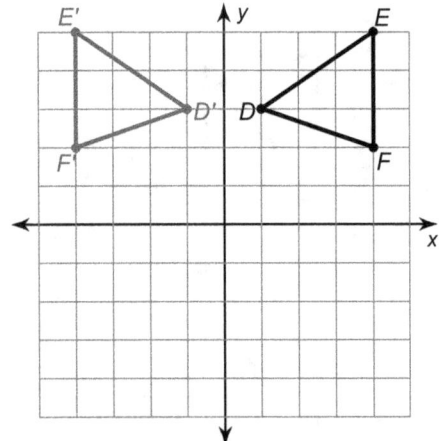

A) Reflection across the x-axis
B) Reflection across y = 5
C) Reflection across the y-axis
D) Reflection across x = 2

12)
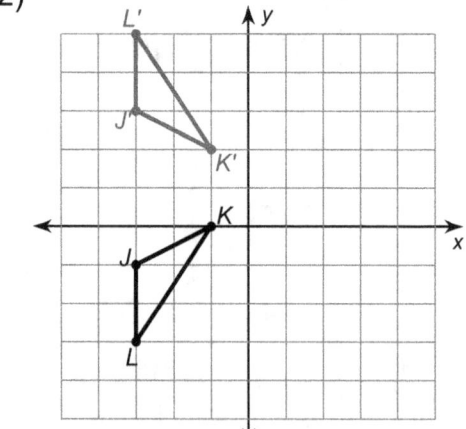

A) Reflection across the y-axis
B) Reflection across y = 1
C) Reflection across x = -1
D) Reflection across the x-axis

Describe the transformation and find the transformation rule :

13)
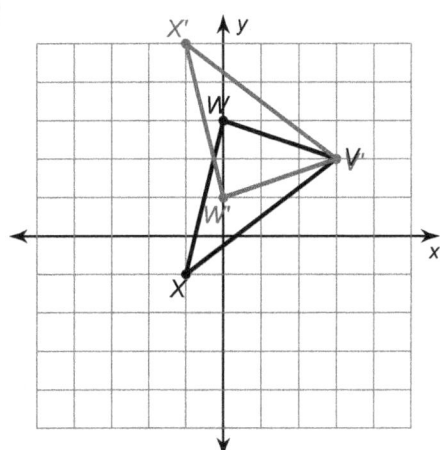

A) Reflection across y = 2
B) Reflection across the x-axis
C) Reflection across x = 2
D) Reflection across y = –2

14)
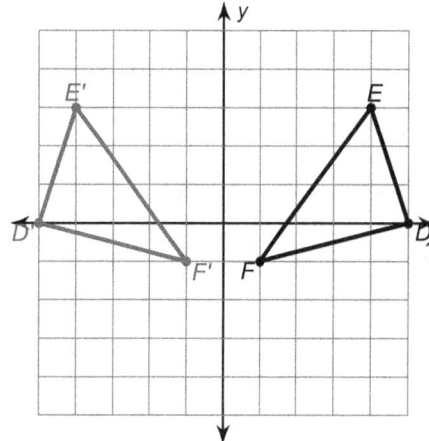

A) Reflection across the y-axis
B) Reflection across x = - 3
C) Reflection across x = - 1
D) Reflection across y = 1

15)
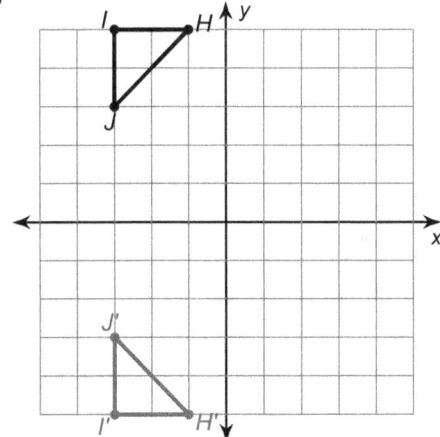

A) Reflection across y = 3
B) Reflection across y = 4
C) Reflection across the y-axis
D) Reflection across the x-axis

Describe the transformation and find the transformation rule :

1)
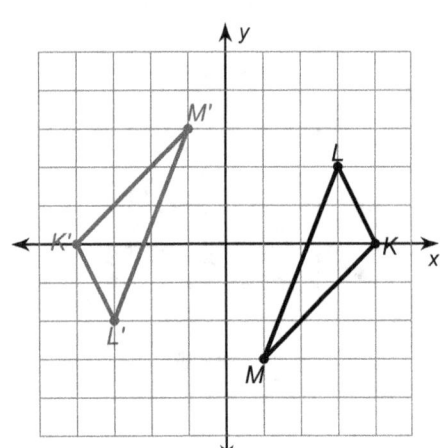

A) Reflection across y = 4
B) Rotation 180° counterclockwise about the origin
C) Rotation 180° about the origin
D) Rotation 90° clockwise about the origin

2)
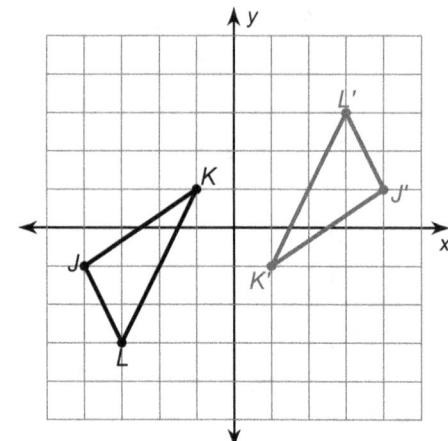

A) Rotation 180° about the origin
B) Rotation 90° clockwise about the origin
C) Rotation 90° counterclockwise about the origin
D) Reflection across x = −1

3)
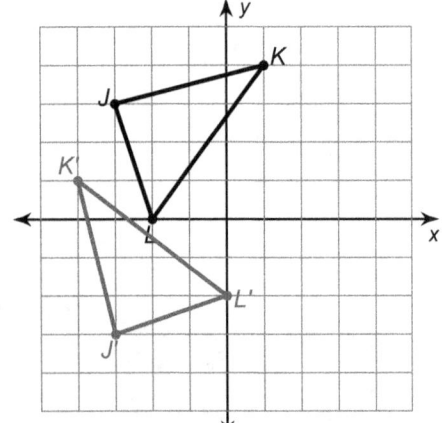

A) Rotation 180° about the origin
B) Rotation 90° clockwise about the origin
C) Rotation 90° counterclockwise about the origin
D) Reflection across y = −2

4)
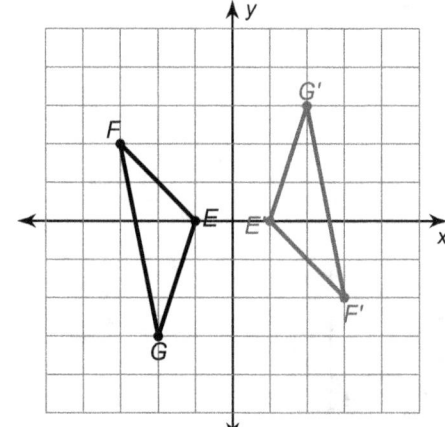

A) Rotation 90° counterclockwise about the origin
B) Rotation 180° about the origin
C) Translation: 6 units left
D) Rotation 90° clockwise about the origin

Describe the transformation and find the transformation rule :

5)

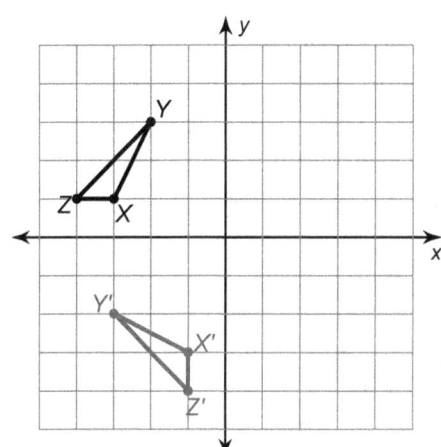

A) Rotation 90° counterclockwise about the origin
B) Rotation 90° clockwise about the origin
C) Translation: 2 units right and 2 units down
D) Rotation 180° about the origin

6)

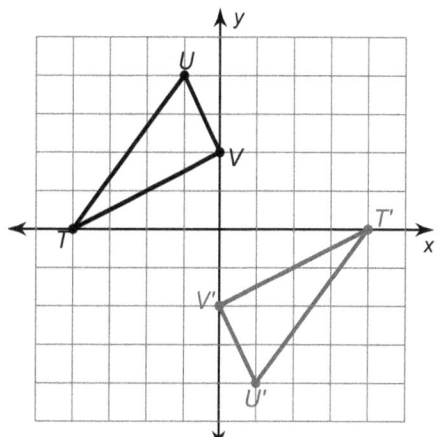

A) Rotation 90° counterclockwise about the origin
B) Rotation 180° about the origin
C) Reflection across y = -2
D) Rotation 90° clockwise about the origin

7)

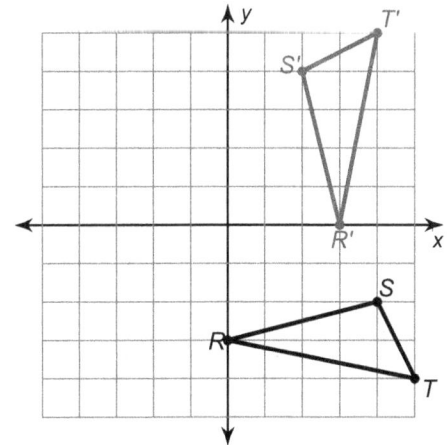

A) Rotation 90° counterclockwise about the origin
B) Rotation 180° about the origin
C) Rotation 90° clockwise about the origin
D) Translation: 5 units down

8)

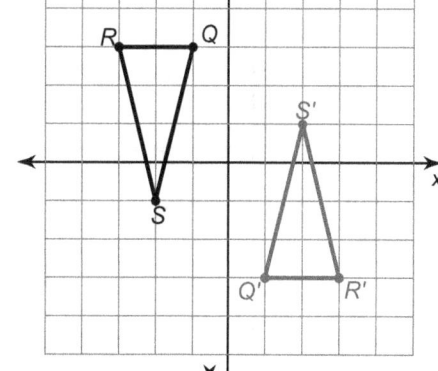

A) Rotation 90° clockwise about the origin
B) Rotation 90° counterclockwise about the origin
C) Translation: 3 units right and 3 units up
D) Rotation 180° about the origin

Describe the transformation and find the transformation rule :

9)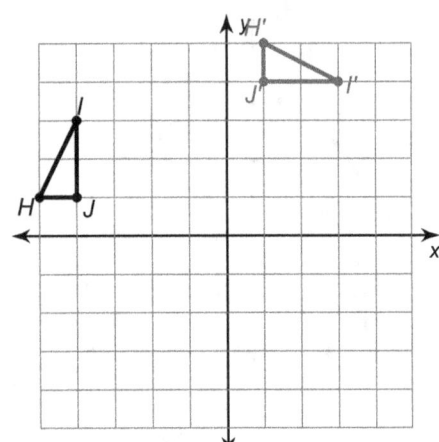

A) Rotation 180° about the origin
B) Rotation 90° counterclockwise about the origin
C) Rotation 90° clockwise about the origin
D) Translation: 7 units down 1 unit up

10)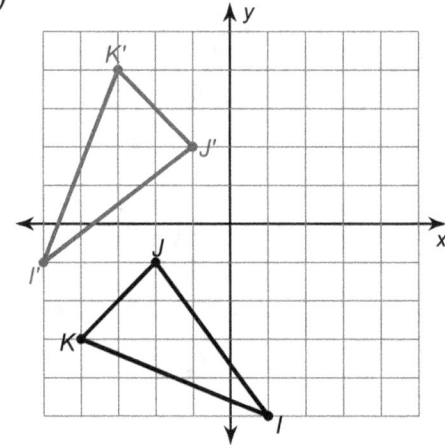

A) Rotation 180° about the origin
B) Rotation 90° clockwise about the origin
C) Reflection across x = 9
D) Rotation 90° counterclockwise about the origin

11)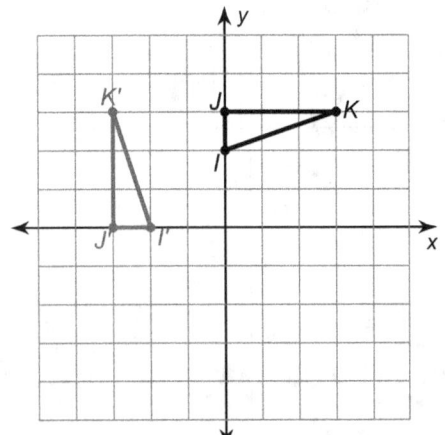

A) Rotation 90° clockwise about the origin
B) Rotation 180° about the origin
C) Reflection across x = -1
D) Rotation 90° counterclockwise about the origin

12)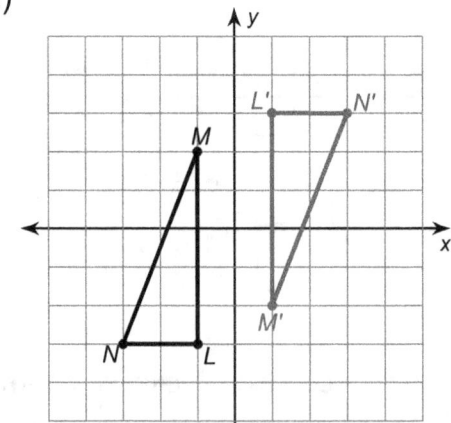

A) Rotation 90° counterclockwise about the origin
B) Rotation 90° clockwise about the origin
C) Rotation 180° about the origin
D) Reflection across the y-axis

Describe the transformation and find the transformation rule:

13)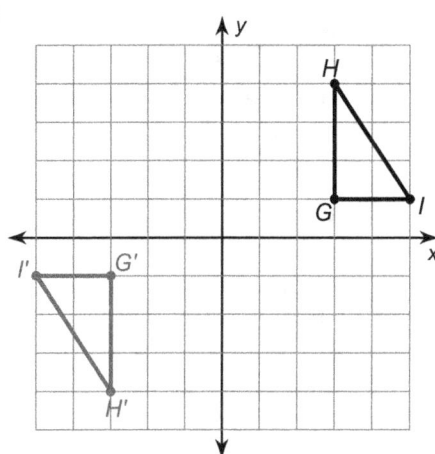

A) Rotation 180° about the origin
B) Reflection across x = 7
C) Rotation 90° counterclockwise about the origin
D) Rotation 90° clockwise about the origin

14)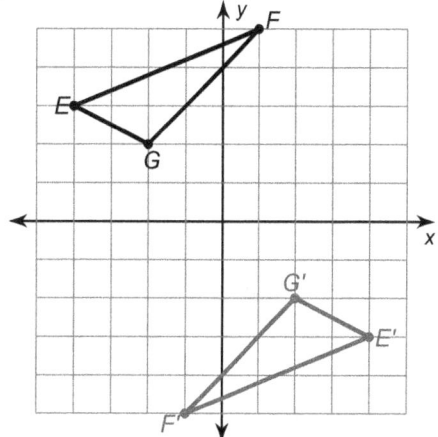

A) Translation: 2 units left 5 units down
B) Rotation 90° clockwise about the origin
C) Rotation 90° counterclockwise about the origin
D) Rotation 180° about the origin

15)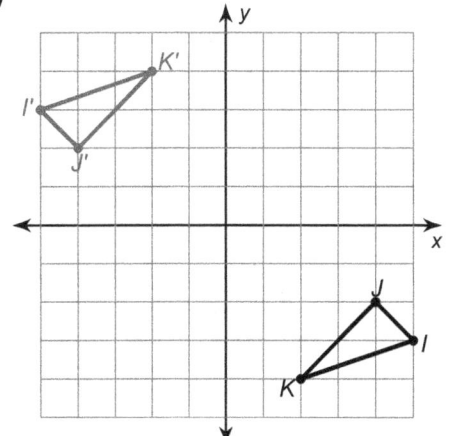

A) Rotation 90° clockwise about the origin
B) Rotation 90° counterclockwise about the origin
C) Reflection across y = −5
D) Rotation 180° about the origin

Translation #56

Describe the transformation and find the transformation rule :

1)

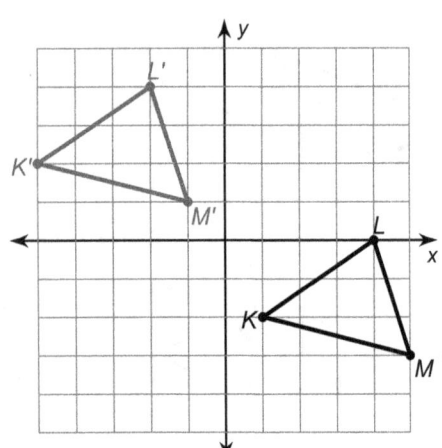

A) Translation: 6 units right and 2 units up
B) Translation: 5 units left and 4 units up
C) Translation: 2 units right
D) Translation: 6 units left and 4 units up

2)

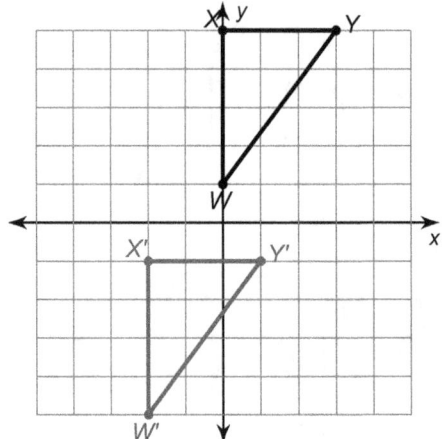

A) Translation: 2 units left and 6 units down
B) Translation: 3 units right and 4 units down
C) Translation: 2 units left and 2 units down
D) Translation: 4 units right and 1 unit up

3)

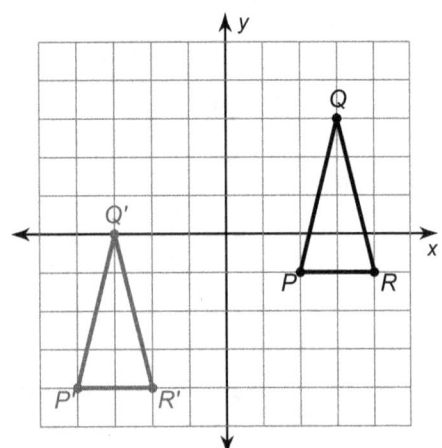

A) Translation: 2 units right and 2 units up
B) Translation: 6 units left and 3 units down
C) Translation: 4 units left and 3 units up
D) Translation: 1 unit left and 1 unit up

4)

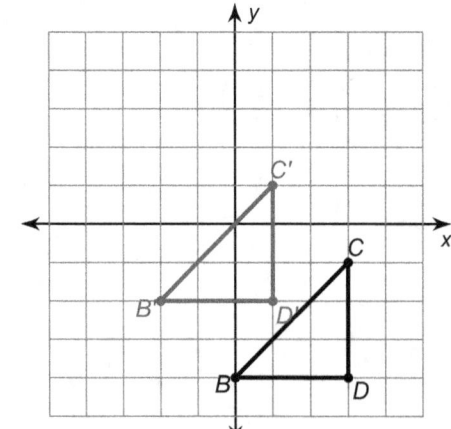

A) Translation: 2 units left and 2 units up
B) Translation: 3 units right and 4 units up
C) Translation: 4 units right and 4 units down
D) Translation: 2 units right and 6 units up

Translation #56

Describe the transformation and find the transformation rule :

5)

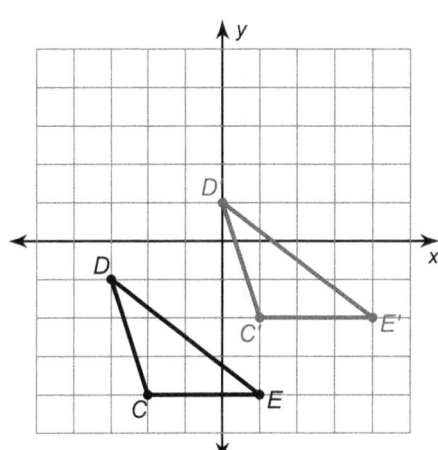

A) Translation: 5 units left and 5 units up
B) Translation: 2 units right and 5 units up
C) Translation: 3 units right and 1 unit down
D) Translation: 3 units right and 2 units up

6)

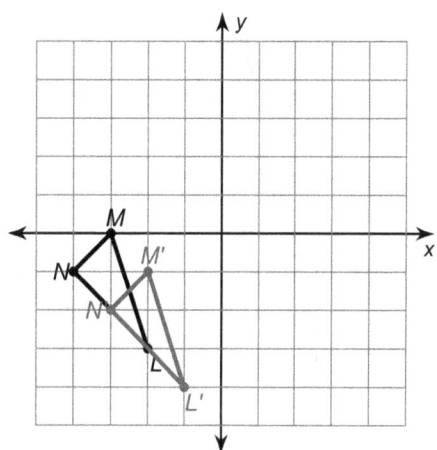

A) Translation: 6 units right and 4 units down
B) Translation: 1 unit right and 1 unit down
C) Translation: 4 units down
D) Translation: 3 units right and 1 unit down

7)

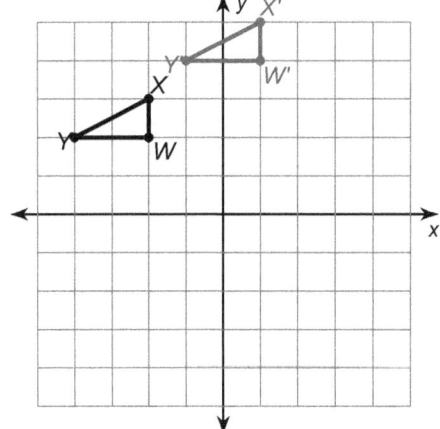

A) Translation: 4 units right and 6 units down
B) Translation: 1 unit right and 4 units up
C) Translation: 5 units right and 4 units down
D) Translation: 3 units right and 2 units up

8)

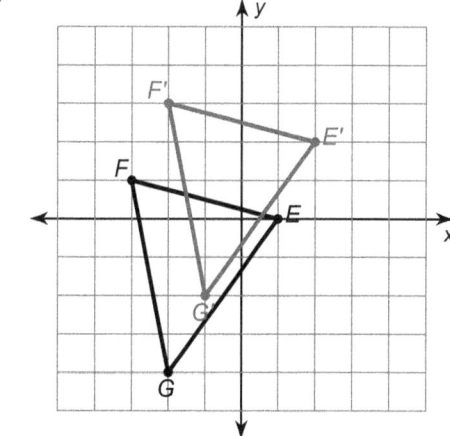

A) Translation: 1 unit right and 3 units down
B) Translation: 1 unit right and 2 units up
C) Translation: 2 units up
D) Translation: 1 unit left and 1 unit up

Describe the transformation and find the transformation rule :

9)

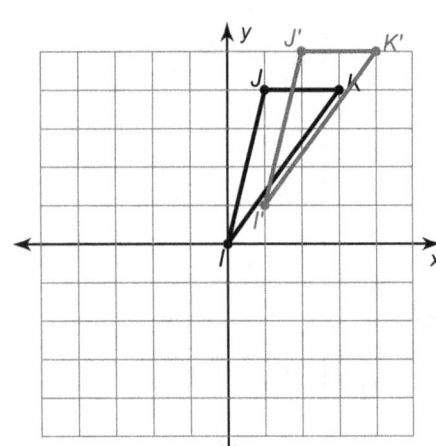

A) Translation: 1 unit right and 1 unit up
B) Translation: 3 units left and 2 units down
C) Translation: 2 unit right and 3 units down
D) Translation: 1 unit left and 1 unit down

10)

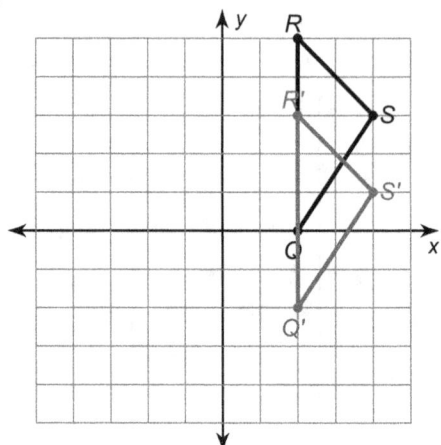

A) Translation: 5 units right and 5 units down
B) Translation: 2 units down
C) Translation: 7 units left and 5 units down
D) Translation: 5 units right and 2 units down

11)

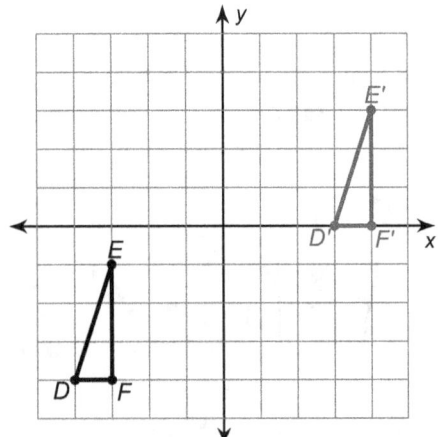

A) Translation: 7 units left and 2 units up
B) Translation: 3 units down
C) Translation: 6 units left and 3 units down
D) Translation: 7 units right and 4 units up

12)

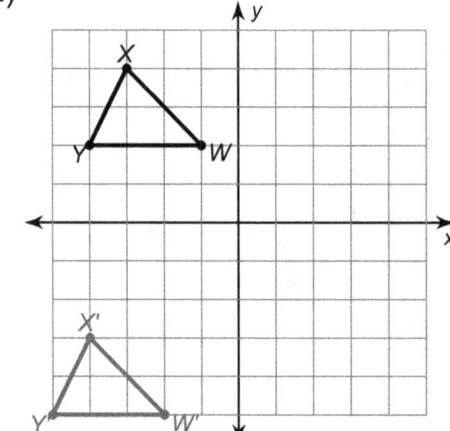

A) Translation: 5 units right and 6 units up
B) Translation: 4 units left and 7 units down
C) Translation: 3 units right and 4 units up
D) Translation: 1 unit left and 7 units down

Describe the transformation and find the transformation rule :

13)

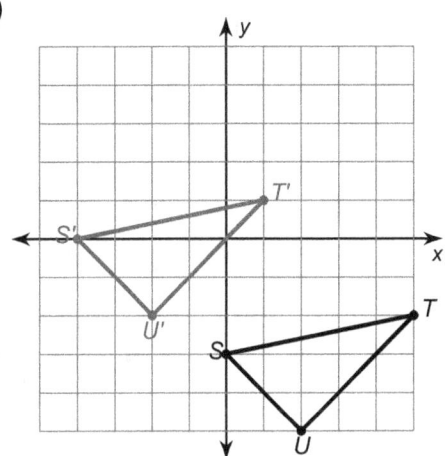

A) Translation: 2 units right
B) Translation: 5 units right and 6 units down
C) Translation: 1 unit left and 1 unit up
D) Translation: 4 units left and 3 units up

14)

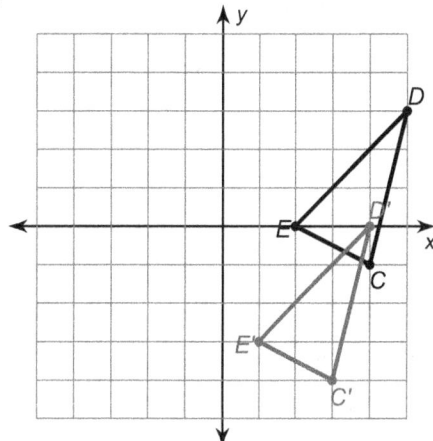

A) Translation: 6 units left and 3 units down
B) Translation: 1 unit left and 3 units down
C) Translation: 4 units right and 4 units up
D) Translation: 2 units left and 2 units down

15)

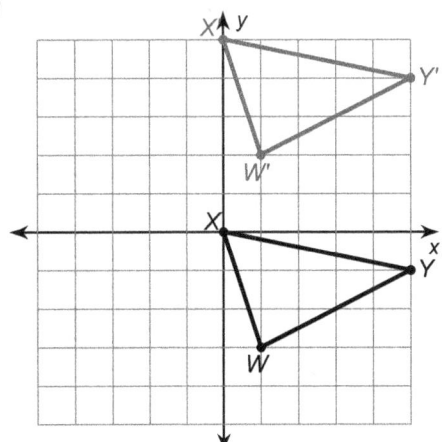

A) Translation: 2 units right and 3 units down
B) Translation: 3 units left
C) Translation: 2 units right and 5 units up
D) Translation: 5 units up

Pre-Algebra Vol 2

ONE-STEP WORD #57

1. Cathy spent $64 to buy a big box of snack bars. If each box cost $8, how many boxes did she buy ?

   (A) 10   (B) 7   (C) 8   (D) 9

2. A cake recipe needs 3 cups of milk. Susan accidentally added 8 cups of milk. How many more cups did she put in ?

   (A) 2   (B) 1   (C) 4   (D) 5

3. Charlie and seven of his friends went for dinner to a restaurant. They split evenly. Each person paid $10. What was the total bill ?

   (A) $70   (B) $86   (C) $80   (D) $1.43

4. Larry ate 15 sugar cookies made by his mom. He ate $\frac{3}{10}$ of the cookies made by his mom. How many cookies are left ?

   (A) 4.5   (B) 35   (C) 58   (D) 50

5. Macy shared 10 of her muffins with her friends. If she shared $\frac{5}{6}$ of what she had. How many muffins are left ?

   (A) 3   (B) 5   (C) 1   (D) 2

6. Jason spent $4.64 for a sandwich at lunch today. She now has $21.20. How much money did she had originally ?

   (A) $25.84　　(B) $22.74　　(C) $30.48　　(D) $16.56

7. Cathy ran 32 miles less than Tony last week. If Cathy ran 17 miles, how many miles did Tony ran ?

   (A) 49　　(B) 15　　(C) 52　　(D) 81

8. Violet has $9 to buy a gift for her friend. She has $\frac{3}{10}$ of the cost of the gift. How much more money she needs to save ?

   (A) $33　　(B) $30　　(C) $36　　(D) $21

9. Dan spent $8.84 for dinner. He now has $16.75. How much money did he had originally with ?

   (A) $26.87　　(B) $7.91　　(C) $25.59　　(D) $34.43

10. Last week Sam ran 22 miles more than Zoya. Sam ran 40 miles last week. How many miles did Zoya ran ?

    (A) 15　　(B) 62　　(C) 18　　(D) 4

11. Olivia is baking cupcakes. The recipe needs $4\frac{3}{4}$ cups of flour. She accidentally pours 8 cups of flour. How many extra cups of flour did she put in?

(A) $12\frac{3}{4}$ (B) $3\frac{1}{4}$ (C) 8 (D) $\frac{19}{32}$

12. Last week Leo ran 8 miles less than Nathan. Leo ran 12 miles last week. How many miles did Nathan ran?

(A) 18 (B) 20 (C) 4 (D) 28

13. Noah and his best friend got a cash prize. They split the money evenly, each getting $23.76. How much money was the cash prize?

(A) $47.52 (B) $48.17 (C) $11.88 (D) $51.80

14. Ava has $5 savings. She wants to buy a bracelet and her savings cover $\frac{1}{4}$ of its costs. How much did the bracelet cost?

(A) $30 (B) $25 (C) $15 (D) $20

15. Charlie won 35 stickers by playing the bean bag toss game at a local fair. He shared some of his stickers with his friends. He is left with 19 stickers. How many stickers did he share with his friends?

(A) 16 (B) 21 (C) 18 (D) 14

A farmhouse has a total of 25 animals containing of horses and chickens. Altogether there are 90 legs.

1. How many chickens are there in the farm ?

(A) 5 chickens (B) 25 chickens

(C) 20 chickens (D) 23 chickens

2. How many horses are there in the farm ?

(A) 20 horses (B) 3 horses

(C) 4 horses (D) 2 horses

A farmhouse has a total of 21 animals containing of geese and cows. Altogether there are 70 legs.

3. How many geese are there in the farm ?

(A) 18 geese (B) 17 geese

(C) 7 geese (D) 19 geese

4. How many cows are there in the farm ?

(A) 3 cows (B) 5 cows

(C) 2 cows (D) 14 cows

A farmhouse has a total of 12 animals containing of buffalo and chickens. Altogether there are 32 legs.

5. How many buffalo are there in the farm ?

(A) 2 buffalo   (B) 6 buffalo
(C) 3 buffalo   (D) 4 buffalo

6. How many chickens are there in the farm ?

(A) 10 chickens   (B) 8 chickens
(C) 9 chickens   (D) 11 chickens

A farmhouse has a total of 17 animals containing of buffalo and chickens. Altogether there are 60 legs.

7. How many chickens are there in the farm ?

(A) 14 chickens   (B) 4 chickens
(C) 13 chickens   (D) 15 chickens

8. How many buffalo are there in the farm ?

(A) 13 buffalo   (B) 3 buffalo
(C) 4 buffalo   (D) 2 buffalo

SYSTEM OF EQUATIONS WORD PROBLEMS #58

A farmhouse has a total of 12 animals containing of ducks and horses. Altogether there are 38 legs.

9. How many ducks are there in the farm ?

(A) 8 ducks (B) 9 ducks

(C) 5 ducks (D) 10 ducks

10. How many horses are there in the farm ?

(A) 4 horses (B) 3 horses

(C) 2 horses (D) 7 horses

A farmhouse has a total of 16 animals containing of ducks and buffalo. Altogether there are 44 legs.

11. How many ducks are there in the farm ?

(A) 10 ducks (B) 13 ducks

(C) 12 ducks (D) 14 ducks

12. How many buffalo are there in the farm ?

(A) 3 buffaloes (B) 6 buffaloes

(C) 4 buffaloes (D) 2 buffaloes

SYSTEM OF EQUATIONS WORD PROBLEMS #58

A farmhouse has a total of 20 animals containing of geese and oxen. Altogether there are 60 legs.

13. How many geese are there in the farm ?

(A) 18 geese

(B) 15 geese

(C) 17 geese

(D) 10 geese

14. How many oxen are there in the farm ?

(A) 12 oxen

(B) 3 oxen

(C) 10 oxen

(D) 2 oxen

15. A farmhouse has a total of 20 animals containing of geese and horses. Altogether there are 72 legs. How many horses and how many geese are there ?

(A) 20 geese and 4 horses

(B) 4 geese and 16 horses

(C) 18 geese and 2 horses

(D) 20 geese and 2 horses

Jack spent $132 on shirts. Plain color shirts cost $28 and striped shirts cost $12. If he bought a total of 7.

1. How many of plain color shirts did he buy ?

(A) 5 (B) 3
(C) 2 (D) 4

2. How many of striped shirts did he buy ?

(A) 2 (B) 4
(C) 5 (D) 3

Cathy bought 9 pairs of dresses for a total of $780. Formal dress cost is $90 and casual dress cost is $80.

3. How many of formal dress did she buy ?

(A) 7 (B) 5
(C) 5 (D) 6

4. How many of casual dress did she buy ?

(A) 2 (B) 3
(C) 4 (D) 5

Daniel spent $28 on cutlery for his birthday party. Cost of each spoon is $5 and each fork cost is $6. He bought a total of 5.

5. How many of spoons did he buy ?

(A) 2 spoons (B) 3 spoons
(C) 1 spoons (D) 4 spoons

6. How many of forks did he buy ?

(A) 4 forks (B) 5 forks
(C) 1 forks (D) 3 forks

Amanda bought 8 bracelets for a total of $84. Plain bracelets cost is $7 and fancy bracelets cost is $14.

7. How many of Plain bracelets did she buy ?

(A) 6 (B) 8
(C) 11 (D) 4

8. How many of fancy bracelets did she buy ?

(A) 9 (B) 2
(C) 4 (D) 3

Hanna bought a total of 9 pens and pencils and paid $15. Each pens cost is $3 and pencils cost is $1.

9. How many of pens did she buy ?

(A) 3 pens (B) 6 pens
(C) 5 pens (D) 7 pens

10. How many of pencils did she buy ?

(A) 3 pencils (B) 6 pencils
(C) 4 pencils (D) 2 pencils

Amanda bought 8 bags for a total of $194. Big bags cost is $29 and small bag cost is $10.

11. How many of Big bags did she buy ?

(A) 13 (B) 4
(C) 8   (D) 6

12. How many of small bags did she buy ?

(A) 4 (B) 2
(C) 7 (D) 3

Olivia spent $192 on puzzles. Puzzle A cost is $21 and puzzle B cost is $25. She bought a total of 8 puzzles.

13. How many of Puzzle A did she buy ?

(A) 6   (B) 4
(C) 5   (D) 2

14. How many of Puzzle B did she buy ?

(A) 2   (B) 3
(C) 3   (D) 6

15. Tony bought 9 toy cars for a total of $91. Red car cost is $4 and blue car cost is $15. How many red and blue cars did he buy ?

(A) 4 red cars and 5 blue cars
(B) 6 red cars and 3 blue cars
(C) 7 red cars and 2 blue cars
(D) 5 red cars and 2 blue cars

Line Graph #60

Based on the below graph answer questions from 1 - 7

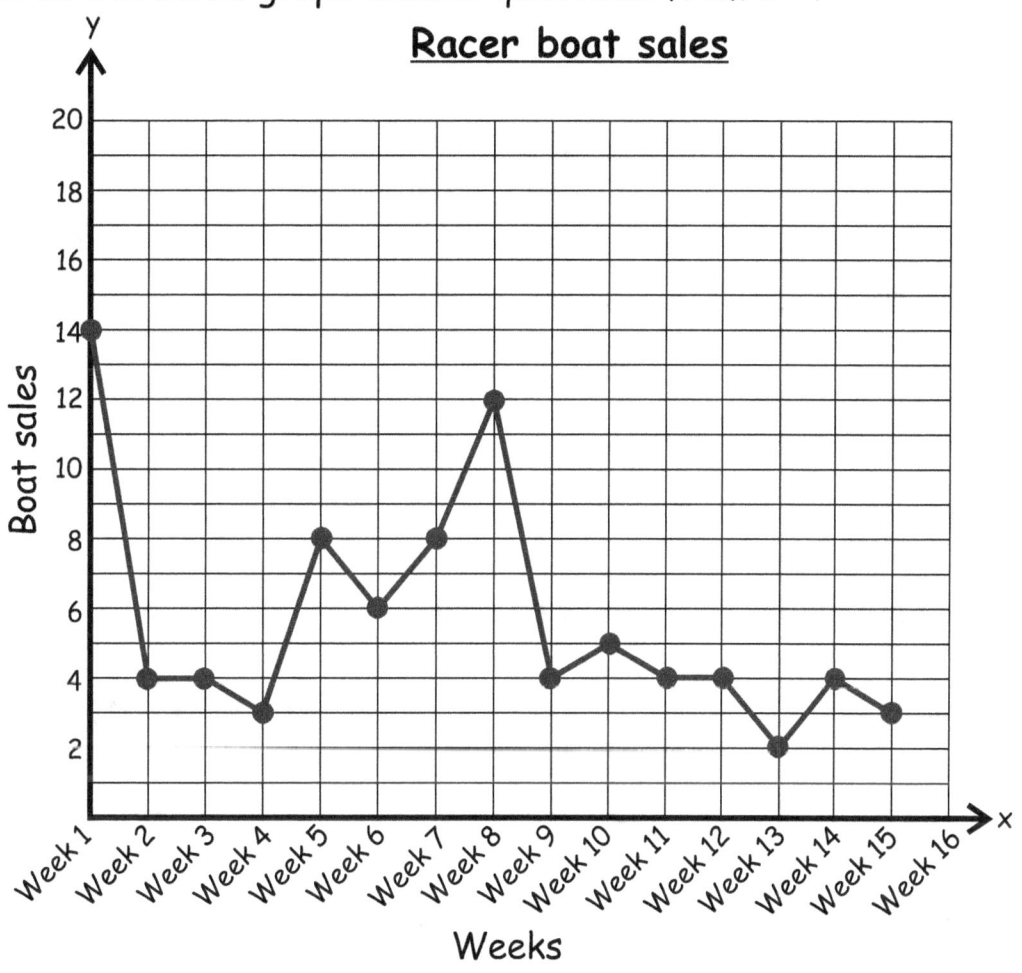

1. Find the median of the above data ?

   (A) 11   (B) 7

   (C) 4   (D) 9.5

2. Find the Mean of the above data ?

   (A) 5.67   (B) 8.39

   (C) 8.23   (D) 9.88

Pre-Algebra Vol 2

Line Graph #60

3. Find the Range of the above data?

(A) 16 (B) 15

(C) 14 (D) 12

4. What is the Mode of the above data?

(A) 4 (B) 11

(C) 7 (D) 9

5. Which week has the highest boat sales?

(A) week 3 (B) week 11

(C) week 1 (D) week 8

6. Which week has the lowest boat sales?

(A) week 13 (B) week 3

(C) week 10 (D) week 8

7. In which week exactly six boats were sold?

(A) week 8 (B) week 9

(C) week 2 (D) week 6

Based on the below graph answer questions from 8 - 14

Profit on video game sales

8. Find the median of the above data ?

(A) 7 (B) 3

(C) 9 (D) 1

9. Find the Mean of the above data ?

(A) 5.66 (B) 5.34

(C) 4.44 (D) 5.3125

10. Find the Range of the above data ?

   (A) 19 (B) 18

   (C) 15 (D) 27

11. What is the Mode of the above data ?

   (A) 5 (B) 7

   (C) 2 and 3 (D) 8

12. Which week has the highest video game sales ?

   (A) week 1 (B) week 13

   (C) week 9 (D) week 16

13. Which week has the lowest video game sales ?

   (A) week 4 (B) week 14

   (C) week 8 (D) week 11

14. In which week exactly six video games were sold ?

   (A) week 11 (B) week 13

   (C) week 12 (D) week 10

Line Graph #60

Based on the below graph answer questions from 15 - 21

15. Find the median of the above data ?

   (A) 16.5   (B) 16.06

   (C) 15.65   (D) 15.95

16. Find the Mean of the above data ?

   (A) 14.39   (B) 16.06

   (C) 15.88   (D) 15.99

17. Find the Range of the above data ?

   (A) 12 (B) 10

   (C) 21 (D) 16

18. What is the Mode of the above data ?

   (A) 19 (B) 21

   (C) 17 (D) 10

19. Which week has the highest sales ?

   (A) week 11 (B) week 14

   (C) week 16 (D) week 8

20. Which week has the lowest sales ?

   (A) week 1 (B) week 9

   (C) week 2 (D) week 13

21. In which week exactly 19 skate boards were sold ?

   (A) week 11 (B) week 15

   (C) week 10 (D) week 8

Based on the below graph answer questions from 22 - 28

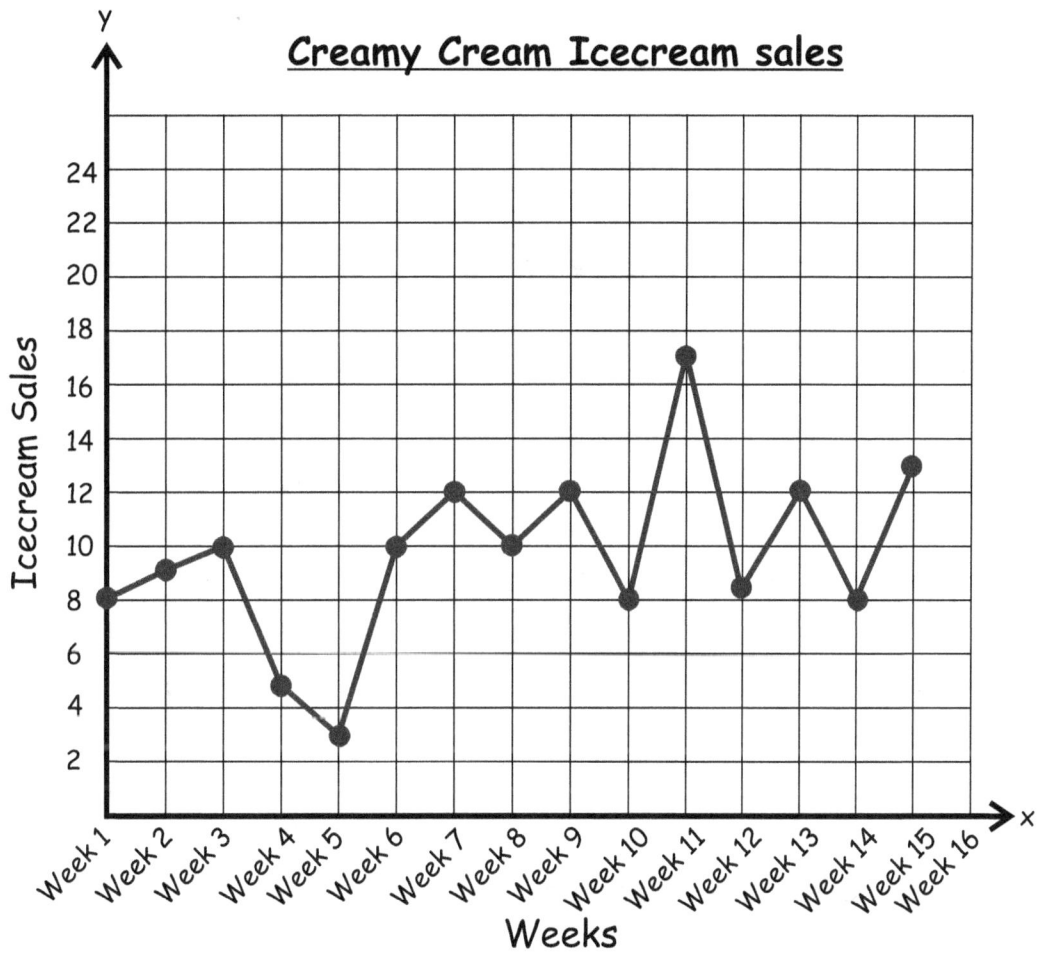

22. Find the median of the above data ?

   (A) 10   (B) 21

   (C) 20   (D) 15

23. Find the Mean of the above data ?

   (A) 9.99   (B) 7.69

   (C) 8.89   (D) 9.67

24. Find the Range of the above data?

   (A) 11   (B) 15

   (C) 8   (D) 14

25. What is the Mode of the above data?

   (A) 10   (B) 8

   (C) 16   (D) 14

26. Which week has the highest sales?

   (A) week 4   (B) week 8

   (C) week 11   (D) week 14

27. Which week has the lowest sales?

   (A) week 1   (B) week 9

   (C) week 5   (D) week 7

28. In which week exactly 13 ice creams were sold?

   (A) week 4   (B) week 8

   (C) week 10   (D) week 15

Based on the below graph answer questions from 29 - 35

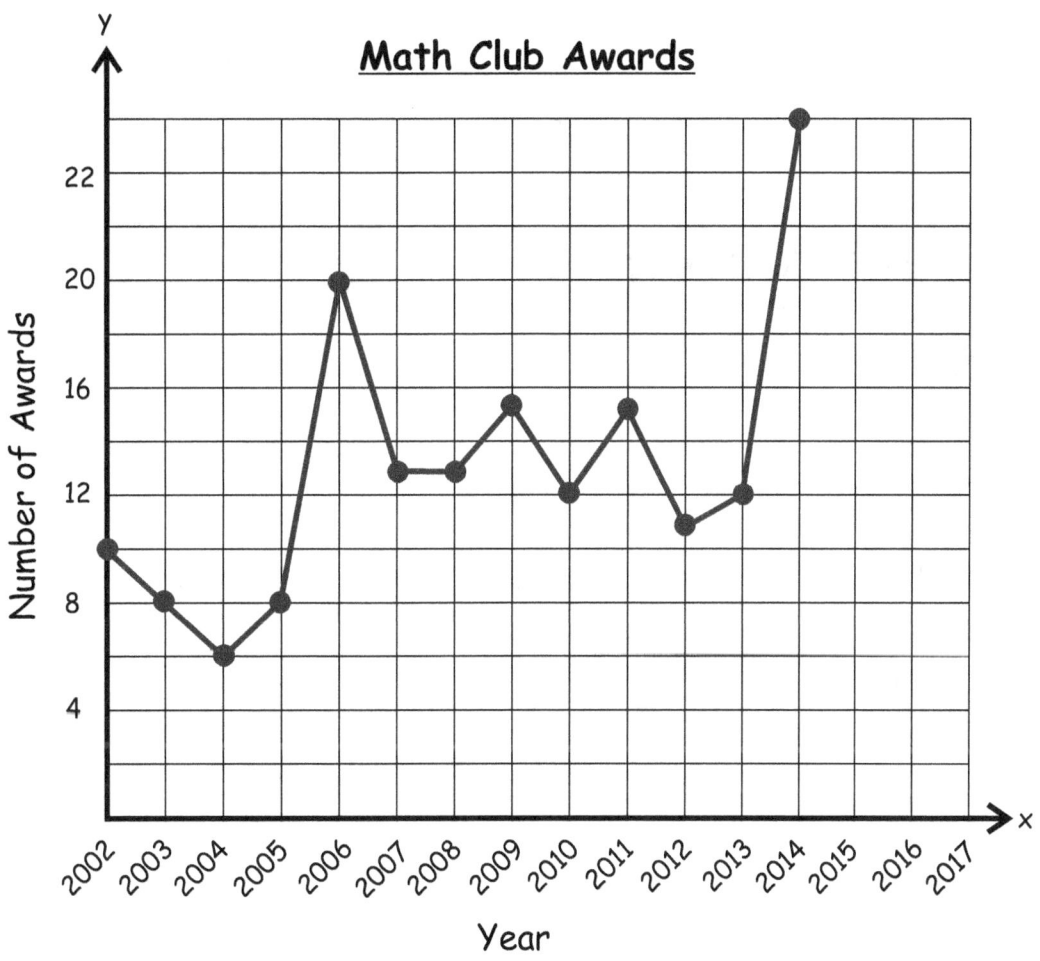

29. Find the median of the above data ?

(A) 12   (B) 10

(C) 16   (D) 16

30. Find the Mean of the above data ?

(A) 15.29   (B) 15.05

(C) 12.99   (D) 12.77

31. Find the Range of the above data ?

    (A) 15   (B) 11

    (C) 17   (D) 19

32. What is the Mode of the above data ?

    (A) 17   (B) 19

    (C) 12,13 and 15   (D) 11

33. During which year math club received highest number of awards ?

    (A) 2004   (B) 2014

    (C) 2011   (D) 2009

34. During which year math club received lowest number of awards ?

    (A) 2012   (B) 2010

    (C) 2004   (D) 2005

35. During which year math club received exactly 20 awards ?

    (A) 2006   (B) 2007

    (C) 2011   (D) 2014

1. Find the correlation between the data plotted as below

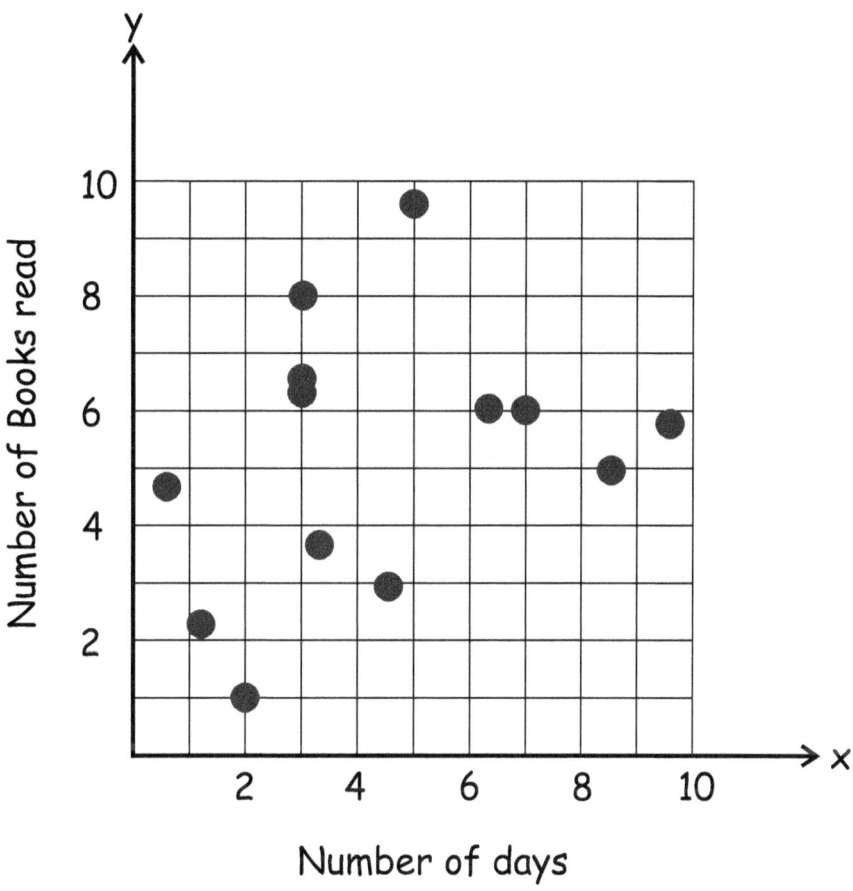

(A) Positive correlation

(B) Negative correlation

(C) No correlation

(D) None

2. Find the correlation between the data plotted as below

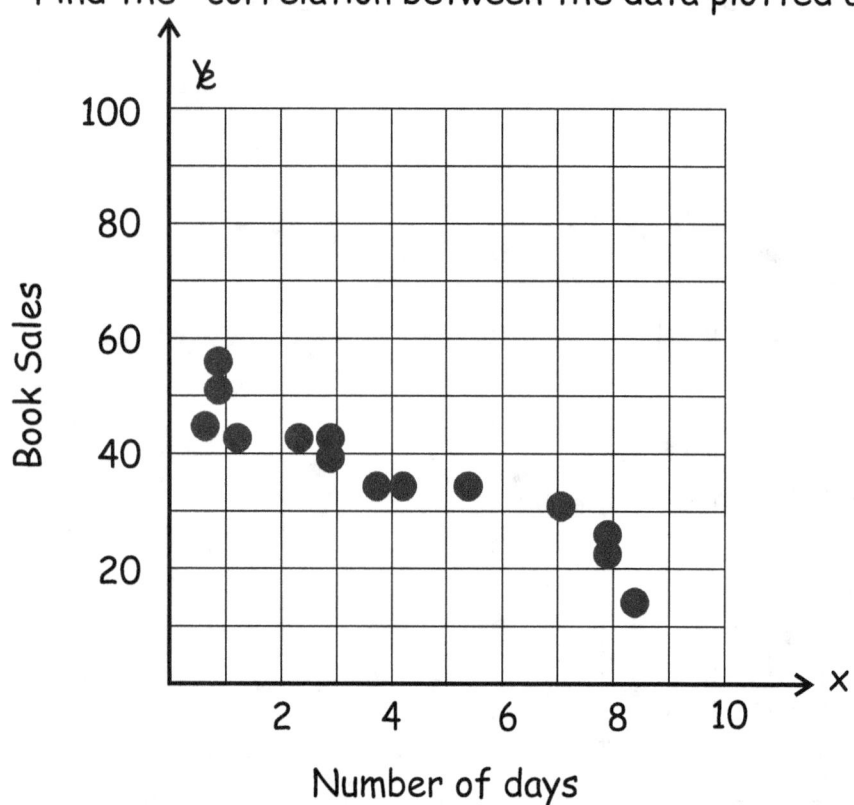

(A) Positive correlation

(B) Negative correlation

(C) No correlation

(D) None

3. Find the correlation between the data plotted as below

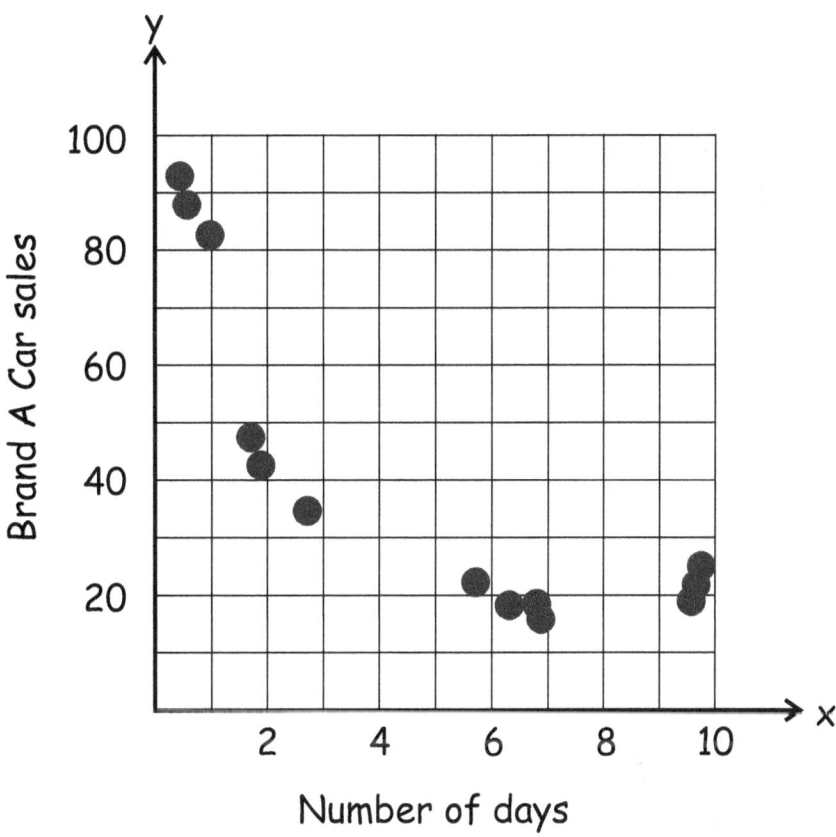

(A) Positive correlation

(B) Negative correlation

(C) No correlation

(D) None

4. Find the correlation between the data plotted as below

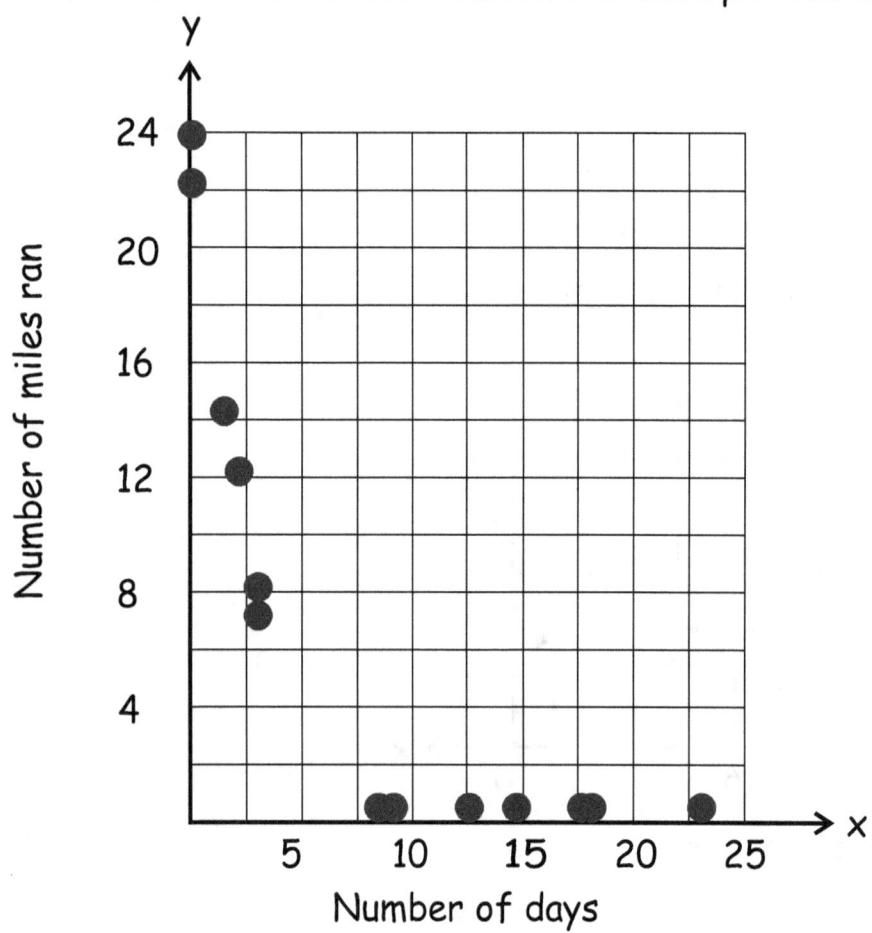

(A) Positive correlation

(B) Negative correlation

(C) No correlation

(D) None

5. Find the correlation between the data plotted as below

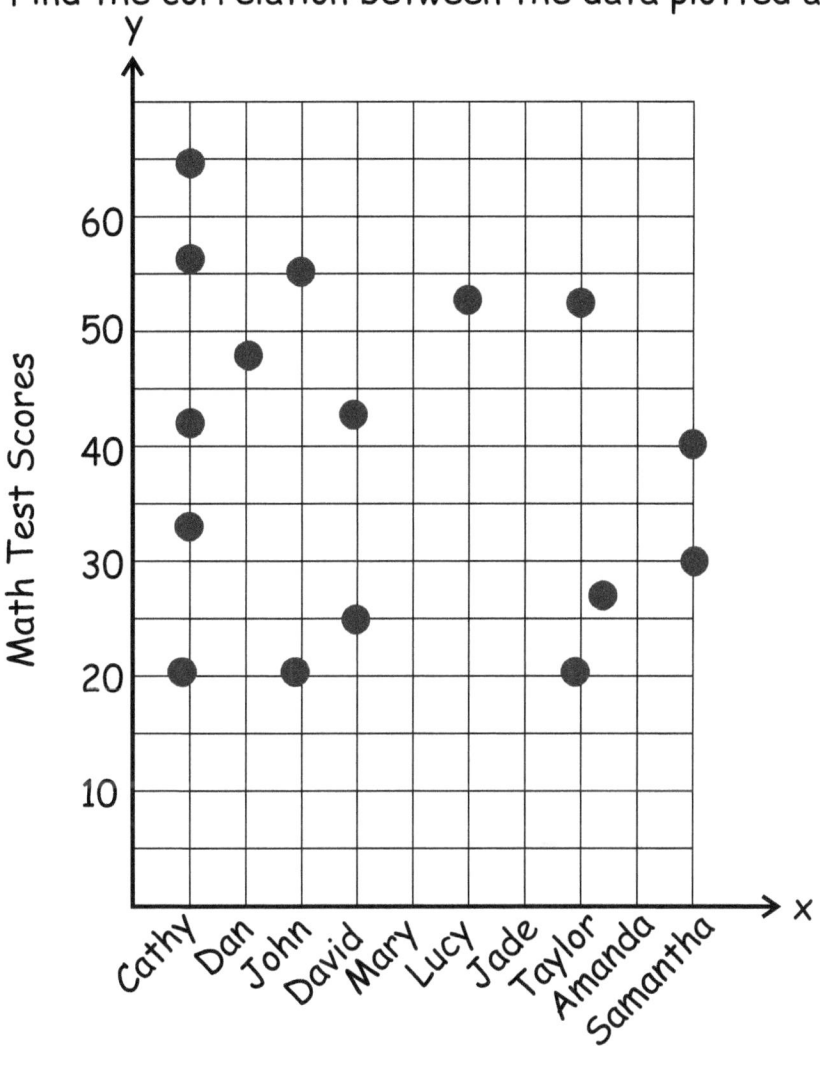

(A) Positive correlation

(B) Negative correlation

(C) No correlation

(D) None

6. Find the correlation between the data plotted as below

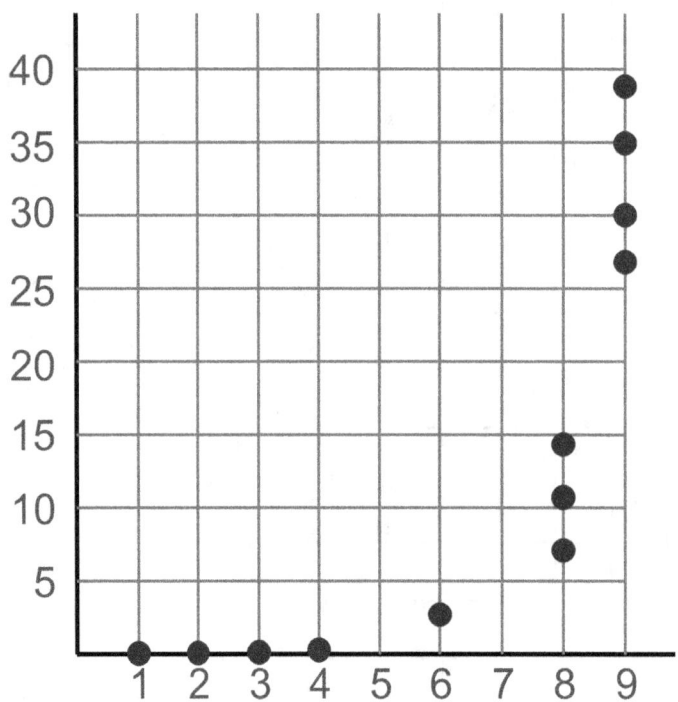

(A) Positive correlation

(B) Negative correlation

(C) No correlation

(D) None

7. Find the correlation between the data plotted as below

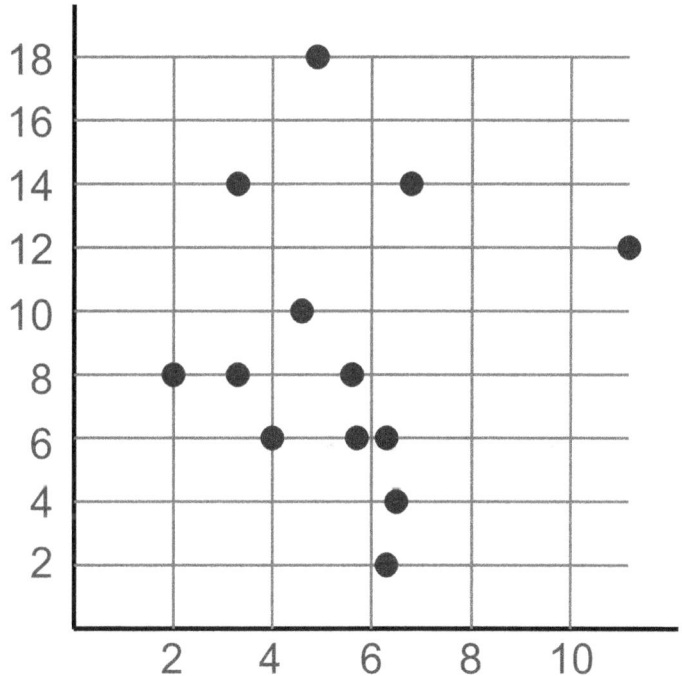

(A) Positive correlation

(B) Negative correlation

(C) No correlation

(D) None

8. Find the correlation between the data plotted as below

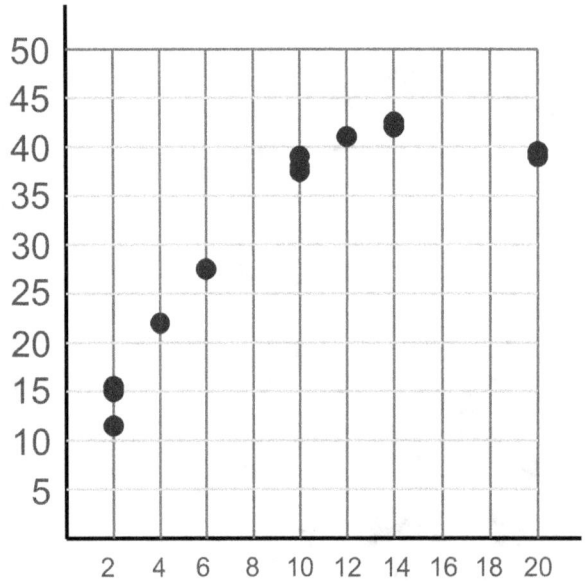

(A) Positive correlation

(B) Negative correlation

(C) No correlation

(D) None

9. Find the correlation between the data plotted as below

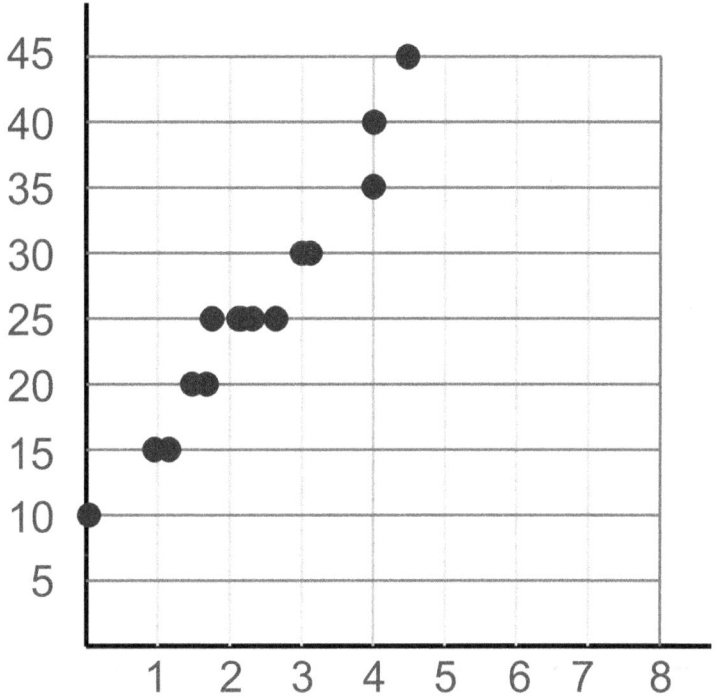

(A) Positive correlation

(B) Negative correlation

(C) No correlation

(D) None

10. Find the correlation between the data plotted as below

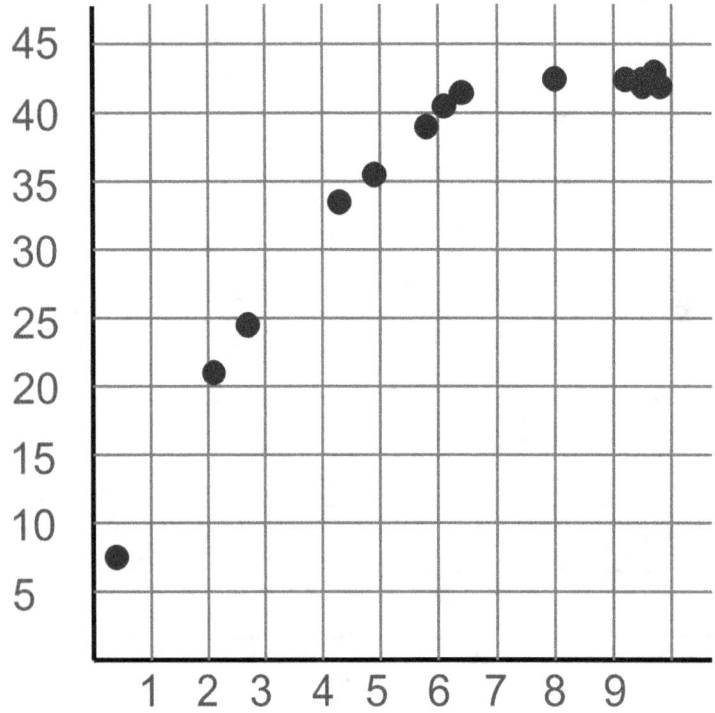

(A) Positive correlation

(B) Negative correlation

(C) No correlation

(D) None

Pre-Algebra Vol 2

Stem Leaf Plot #62

Girls Scout team raised funds through various lemonade stands across the community. Number of lemonade glasses sold are plotted as below. Answer the questions 1 - 4.

Stem	Leaf
1	7 9
2	5 5 5 6 6 7 8 9 9
3	1 1 3 8
4	9

0 | 1 = 1 Glass

1. Find the median of the above data ?

   (A) 28.5   (B) 26.5

   (C) 27.5   (D) 15.9

2. Find the Mean of the above data ?

   (A) 28.63   (B) 25.57

   (C) 31.33   (D) 22.43

3. Find the Range of the above data ?

   (A) 18   (B) 37

   (C) 65   (D) 32

Pre-Algebra Vol 2

Stem Leaf Plot #62

4. What is the Mode of the above data?

   (A) 25   (B) 31

   (C) 29   (D) 35

Larry sold a number of town homes over the summer. The house prices are listed below. Based on this data answer the questions 5 - 8.

Stem	Leaf
0	1 1 1 1 2 2 2 3
1	2 4 8
2	1 3
3	4
4	3
5	8

20 | 1 = $201 K
       $201,000

5. Find the median of the above data?

   (A) 7.5K   (B) 2.5K

   (C) 3.9K   (D) 7.7K

6. Find the Mean of the above data?

   (A) 15 K   (B) 19 K

   (C) 17.4 K   (D) 14.75 K

Stem Leaf Plot #62

7. Find the Range of the above data?

(A) 61 K   (B) 55 K

(C) 57 K   (D) 47 K

8. What is the Mode of the above data?

(A) 1 K   (B) 1.9 K

(C) 7.5 K   (D) 0.1 K

Grade 6 students took a science quiz and competed for the national science bowl competition. The quiz scores are plotted as below. Based on this data answer this questions 9 - 12.

Stem	Leaf
0	0 1 1 1 1 2 2 3 7 7 7
1	1 1 5 5
2	
3	6

5 | 1 = 51 Points

9. Find the median of the above data?

(A) 7.5   (B) 6.3

(C) 5   (D) 5.9

Pre-Algebra Vol 2

Stem Leaf Plot #62

10. Find the Mean of the above data ?

   (A) 7.6   (B) 9.7

   (C) 8.4   (D) 7.7

11. Find the Range of the above data ?

   (A) 36   (B) 39

   (C) 30   (D) 34

12. What is the Mode of the above data ?

   (A) 7   (B) 11

   (C) 15   (D) 1

A car garage has repaired a number of cars this Quarter. Based on this data answer the questions 13 - 16.

Stem	Leaf
0	8
1	3 3 6
2	3
3	0 2 3 8 9
4	3 7 8

0 | 2 = 2 Cars

13. Find the median of the above data ?

   (A) 32   (B) 44

   (C) 22   (D) 77

14. Find the Mean of the above data ?

   (A) 27.99   (B) 27.76

   (C) 25.46   (D) 29.46

15. Find the Range of the above data ?

   (A) 54   (B) 40

   (C) 39   (D) 46

16. What is the Mode of the above data ?

   (A) 13   (B) 33

   (C) 16   (D) 27

Grade 5 students took math test and the scores are listed as below. Answer the questions 17 - 20

Stem	Leaf
4	9
5	0 2
6	2 2 3 9
7	2 9
8	1 2

5 | 0 = 50 Points

17. Find the median of the above data ?

   (A) 69   (B) 62

   (C) 63   (D) 79

18. Find the Mean of the above data ?

   (A) 65.55   (B) 66.55

   (C) 76.65   (D) 69.05

19. Find the Range of the above data ?

   (A) 42   (B) 35

   (C) 30   (D) 33

20. What is the Mode of the above data ?

   (A) 62   (B) 49

   (C) 69   (D) 81

Dot Plot #63

Super Soccer team played various games across the fall season. The number of goals made are plotted below. Answer the questions 1 - 4.

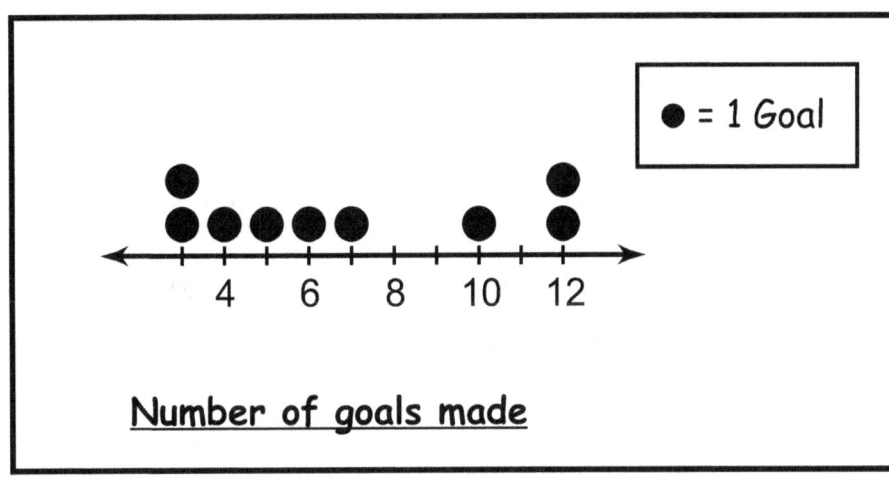

Number of goals made

1. Find the median of the above data ?

   (A) 4   (B) 7

   (C) 6   (D) 5

2. Find the Mean of the above data ?

   (A) 6.89   (B) 6.98

   (C) 6.85   (D) 6.67

3. Find the Range of the above data ?

   (A) 14   (B) 5

   (C) 7   (D) 9

Dot Plot #63

4. What is the Mode of the above data ?

(A) 3 and 12

(B) 5 and 12

(C) 6

(D) 7 and 3

Store A sold a number of cars int he last 11 days as shown below. Based on the data answer the questions 5 - 8.

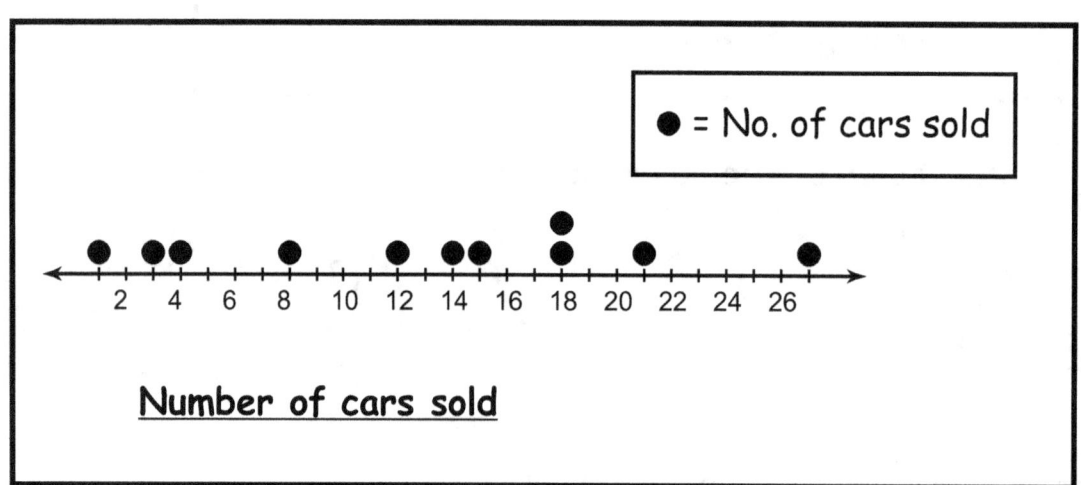

Number of cars sold

5. Find the median of the above data ?

(A) 20

(B) 11

(C) 14

(D) 17

6. Find the Mean of the above data ?

(A) 12.82

(B) 12.33

(C) 12.18

(D) 13.55

7. Find the Range of the above data ?

   (A) 35 (B) 27

   (C) 22 (D) 26

8. What is the Mode of the above data ?

   (A) 23 (B) 18

   (C) 15 (D) 21

Below data gives the average snow fall in Chicago during January 2019. Answer the questions 9 - 12

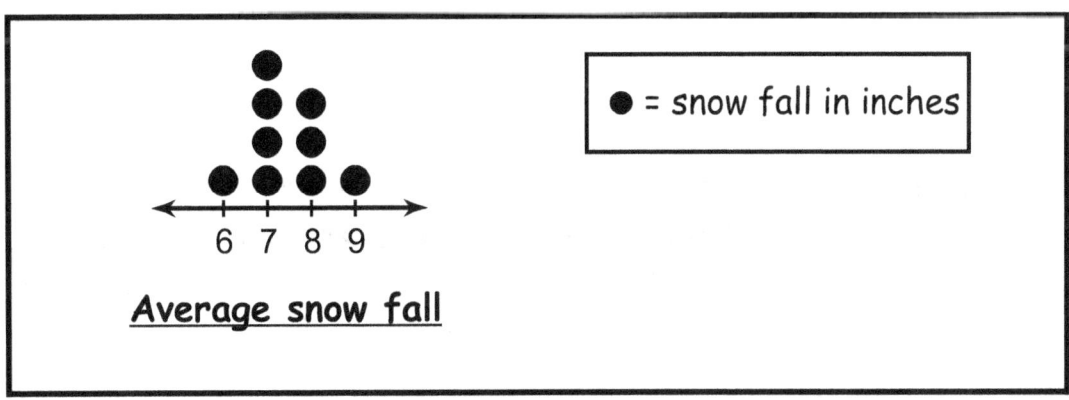

9. Find the median of the above data ?

   (A) 9 (B) 6

   (C) 5 (D) 7

10. Find the Mean of the above data ?

    (A) 7.44 (B) 7.33

    (C) 7.49 (D) 7.95

11. Find the Range of the above data ?

    (A) 3 (B) 11

    (C) 5 (D) 7

12. What is the Mode of the above data ?

    (A) 5 (B) 1

    (C) 12 (D) 7

RC middle school won a number of medals this year in STEM Competitions a plotted below. Answer the questions 13 - 16

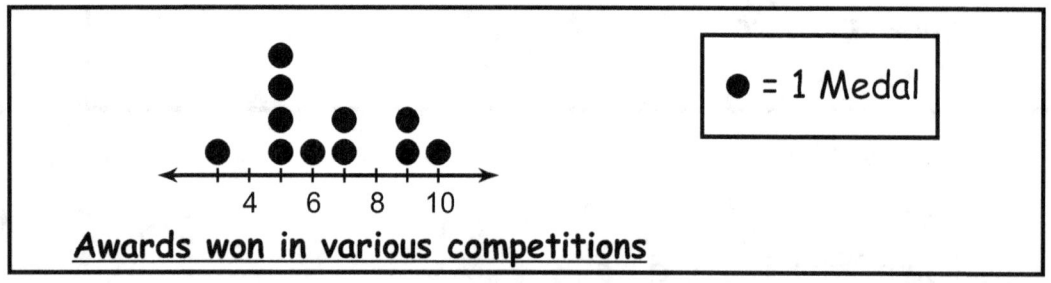

13. Find the median of the above data ?

    (A) 6 (B) 3

    (C) 4 (D) 5

14. Find the Mean of the above data ?

    (A) 6.46   (B) 5.53

    (C) 6.53   (D) 7.63

15. Find the Range of the above data ?

    (A) 8   (B) 12

    (C) 10   (D) 7

16. What is the Mode of the above data ?

    (A) 13   (B) 9

    (C) 5   (D) 2

A robot manufacturing company spends $2880k every year as given in the pie chart below. Answer the questions 1 - 5.

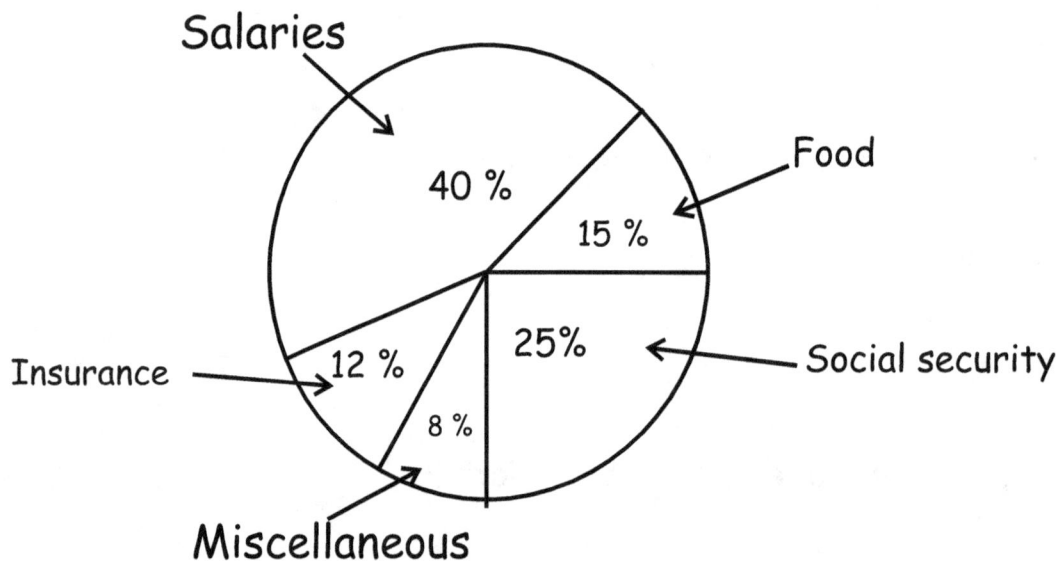

1. Find the amount was spent towards salaries ?

   (A) $1152k　　　　　　　　　　(B) $1252k

   (C) $1100k　　　　　　　　　　(D) $1052k

2. How much money is saved if the budget allocated to miscellaneous is not spent and food went 10% over the budget ?

   (A) $1972k　　　　　　　　　　(B) $187.2k

   (C) $2736k　　　　　　　　　　(D) $2076k

3. How much money was spent on food ?

   (A) 432k   (B) 335k

   (C) 444k   (D) 533k

4. How much money was spent on insurance ?

   (A) 349k   (B) 395.9k

   (C) 312.6k   (D) 345.6k

5. How much more money was spent on social security than insurance ?

   (A) 377.38k   (B) 384.49k

   (C) 374.4k   (D) 399.99k

6. Find the ratio between the amount spent on food and social security ?

   (A) 2:5   (B) 5:3

   (C) 3:5   (D) 2:3

The annual expenditure of an average house hold in the Washington DC is given in the below pie chart. Answer the questions 7 - 12.

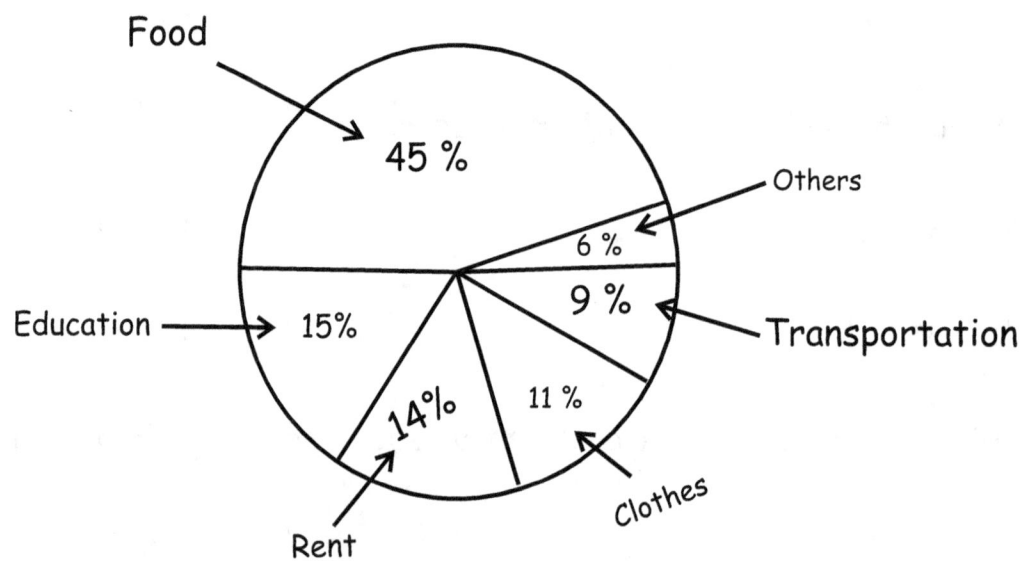

Annual House hold expenditure

7. Find the ratio of expenses made on education and food ?

   (A) 3:1   (B) 1:3

   (C) 2:5   (D) 2:3

8. Find the ratio of expenses made on rent and others ?

   (A) 7:3   (B) 3:7

   (C) 5:7   (D) 7:2

9. Find the ratio of expenses made on clothes and transportation ?

   (A) 1:11   (B) 6:11

   (C) 11:9   (D) 9:11

10. Tony's annual income is $25,000. Find the amount spent on rent and food ?

    (A) $17,250   (B) $16,750

    (C) $11,250   (D) $14,750

11. Tony's annual income is $25,000. Find the amount spent on rent and clothes ?

    (A) $6,250   (B) $6,575

    (C) $6,650   (D) $7,625

12. Tony's annual income is $25,000. Find the amount spent on education ?

    (A) $3,350   (B) $3,880

    (C) $3,745   (D) $3,750

A happy living County people use various transportation methods to go to work. The below pie chart represents the data of 1400 citizens. Answer the questions 13 - 25.

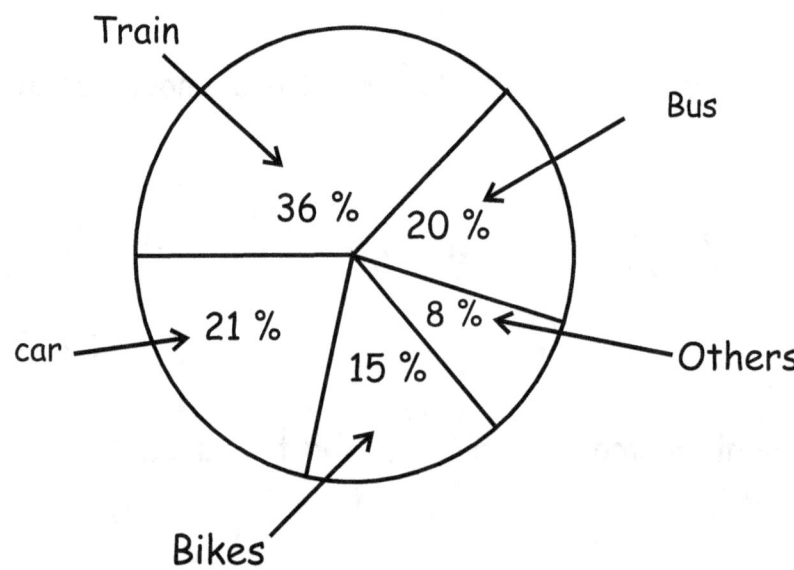

13. Find the ratio of the people travelling in bikes to cars?

    (A) 4:7

    (B) 7:5

    (C) 5:7

    (D) 3:5

14. How many more people travel in train than in cars?

    (A) 210

    (B) 462

    (C) 562

    (D) 452

15. If 1,400 people travel to work everyday, how many people use other as means of transportation ?

    (A) 142   (B) 132

    (C) 112   (D) 122

16. If 1,400 people travel to work everyday, how many people use cars as means of transportation ?

    (A) 394   (B) 304

    (C) 214   (D) 294

17. Find the ratio of the people traveling in trains to buses ?

    (A) 9:5   (B) 12:5

    (C) 5:13   (D) 5:9

18. Find the ratio of the people traveling in bikes to buses ?

    (A) 5:4   (B) 3:5

    (C) 3:4   (D) 4:3

19. Find the ratio of the people traveling in cars to others ?

    (A) 7:4   (B) 21:8

    (C) 8:21   (D) 4:7

20. If 1,400 people travel to work everyday, how many people use bikes as means of transportation ?

   (A) 210
   (B) 200
   (C) 225
   (D) 255

21. If 1,400 people travel to work everyday, how many people use buses as means of transportation ?

   (A) 255
   (B) 280
   (C) 288
   (D) 264

22. If 1,400 people travel to work everyday, how many people use train as means of transportation ?

   (A) 409
   (B) 594
   (C) 500
   (D) 504

23. Find the ratio of the people using trains and bikes for transportation ?

   (A) 5:12
   (B) 1:12
   (C) 12:5
   (D) 5:11

24. Find the ratio of the people using bikes and other means for transportation ?

   (A) 15:8
   (B) 8:11
   (C) 11:18
   (D) 8:15

Pre-Algebra Vol 2

Pie Chart #64

25. Find the ratio of the people using buses and cars for transportation ?

   (A) 5:4

   (B) 7:5

   (C) 20:21

   (D) 21:20

Perfect printers company spent their income in various categories. The below pie chart represents the expenses.
Answer the questions 26 - 35

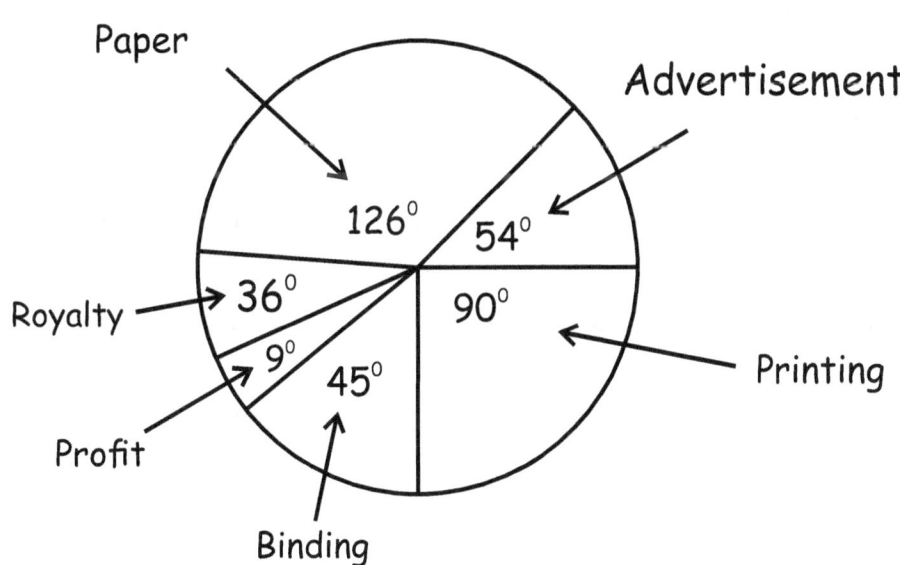

26. Find the ratio of expenses between advertisement and royalty ?

   (A) 2:1

   (B) 2:3

   (C) 3:2

   (D) 1:3

27. Find the ratio of expenses between advertisement and printing ?

   (A)   5:3   (B)   4:7

   (C)   2:5   (D)   3:5

28. Find the ratio of expenses between printing and royalty ?

   (A)   2:5   (B)   5:2

   (C)   1:2   (D)   2:1

29. Find the ratio of expenses between printing and profit ?

   (A)   1:10   (B)   10:11

   (C)   11:10   (D)   10:1

30. Find the ratio of expenses between binding and printing ?

   (A)   2:1   (B)   1:2

   (C)   10:3   (D)   7:5

31. Find the ratio of expenses between printing and paper ?

   (A)   7:5   (B)   10:3

   (C)   5:7   (D)   3:10

32. Find the ratio of expenses between royalty and profit ?

   (A) 4:1   (B) 1:4

   (C) 4:7   (D) 7:4

33. Find the ratio of expenses between binding and profit ?

   (A) 5:1   (B) 4:5

   (C) 7:4   (D) 1:5

34. Find the ratio of expenses between binding and paper ?

   (A) 5:7   (B) 7:5

   (C) 5:14   (D) 14:5

35. Find the ratio of expenses between profit and advertisement ?

   (A) 6:1   (B) 7:6

   (C) 6:7   (D) 1:6

Pre-Algebra Vol 2

KEY

Q #	#1 Fractions Multiply	#2 Fractions Divide	#3 Fractions Subtractions	#4 Decimals Add_Subtractions
1	C	B	B	D
2	D	A	C	D
3	C	C	A	A
4	C	D	A	D
5	B	A	C	A
6	D	C	B	B
7	A	C	C	B
8	A	A	D	B
9	B	A	C	B
10	C	B	A	B
11	C	A	D	B
12	A	D	D	D
13	D	B	B	A
14	C	B	C	D
15	A	C	C	B

Pre-Algebra Vol 2

KEY

Q #	#5 Decimal Multiplication	#6 Decimal Divisions	#7 Integer Multiplication	#8 Scientific Notation
1	D	A	B	C
2	D	C	C	D
3	A	C	A	B
4	D	B	A	D
5	D	D	B	C
6	C	A	A	B
7	C	B	A	B
8	D	A	C	D
9	D	B	D	C
10	D	C	D	C
11	D	B	B	D
12	A	D	C	D
13	B	C	B	C
14	C	D	A	D
15	C	D	D	B

Q #	#9 GCF	#10 GCF Monomials	#11 LCM Numbers	#12 LCM Monomials
1	C	D	B	A
2	D	B	D	B
3	C	A	B	C
4	B	A	C	A
5	C	D	C	C
6	A	D	B	C
7	D	D	C	D
8	B	B	B	C
9	D	A	D	B
10	A	D	A	C
11	A	A	B	A
12	A	B	C	A
13	D	A	D	D
14	C	A	A	C
15	A	D	B	D

KEY

Q #	#13 Order of Operations	#14 Verbal Expressions	#15 VerbalExpression Equation	#16 Monomials
1	D	A	A	D
2	C	D	D	C
3	D	A	D	C
4	C	A	D	C
5	A	B	D	A
6	A	C	C	D
7	D	A	C	B
8	D	B	D	A
9	B	D	C	C
10	C	B	B	C
11	D	D	B	D
12	B	A	D	C
13	C	A	D	B
14	A	D	D	C
15	C	A	A	D

Pre-Algebra Vol 2

KEY

Q #	#17 Inequalities	#18 Evaluate Expressions	#19 Evaluate Expressions2	#20 Solve for X
1	D	B	A	B
2	C	D	A	C
3	D	B	B	D
4	B	B	D	A
5	A	C	A	D
6	C	C	A	A
7	A	B	C	A
8	D	A	C	B
9	C	D	B	B
10	A	B	D	A
11	C	A	A	A
12	B	B	B	C
13	C	A	B	B
14	C	B	D	A
15	B	A	B	B

Pre-Algebra Vol 2

KEY

Q #	#21 Absolute Value	#22 Absolute Value	#23 Proportions	#24 Proportions
1	B	A	A	B
2	A	B	B	C
3	D	B	A	B
4	B	D	A	A
5	A	B	C	B
6	B	C	B	C
7	D	B	C	C
8	A	C	A	A
9	B	A	D	D
10	C	C	D	B
11	A	D	B	C
12	C	D	A	A
13	D	A	D	C
14	A	C	A	A
15	A	B	A	B

KEY

Q #	#25 Percent Discount	#26 Percent MarkUp	#27 Percent Tax	#28 Percent Change
1	A	A	D	A
2	C	B	B	C
3	D	B	D	D
4	D	B	D	D
5	C	A	C	C
6	A	B	D	A
7	C	C	B	A
8	A	C	D	C
9	B	B	A	C
10	D	A	B	A
11	A	A	A	D
12	C	D	C	B
13	D	D	B	B
14	D	A	B	D
15	D	D	D	D

KEY

Q #	#29 Radicals	#30 Radicals	#31 Radicals	#32 Radicals
1	C	B	A	C
2	C	D	A	A
3	B	B	B	C
4	C	B	C	A
5	A	C	D	D
6	C	B	D	A
7	B	B	C	B
8	C	D	D	C
9	C	B	B	C
10	D	C	A	C
11	D	B	D	C
12	C	B	C	C
13	C	C	A	B
14	A	B	A	D
15	B	D	D	D

KEY

Q #	#33 Radicals	#34 Radicals	#35 Triangles	#36 Rectangle square
1	A	A	A	B
2	B	B	C	C
3	D	B	D	B
4	B	C	B	D
5	C	D	D	A
6	D	D	A	D
7	D	C	C	D
8	D	D	D	A
9	B	B	B	C
10	D	A	A	D
11	A	D	D	B
12	B	B	C	C
13	C	B	C	A
14	D	D	C	C
15	B	C	A	A

Pre-Algebra Vol 2

KEY

Q #	#37 Parallelogram	#38 Circle-Area-Circumference	#39 Volume-Sphere	#40 Volume-Rectangle-Square
1	B	A	D	D
2	A	D	D	A
3	A	C	C	C
4	B	A	A	C
5	C	B	A	C
6	D	B	D	A
7	D	A	A	A
8	C	B	C	C
9	D	A	A	C
10	C	C	C	A
11	D	C	C	D
12	A	C	A	B
13	C	A	B	D
14	C	D	B	C
15	D	A	A	A

KEY

Q #	#41 Volume - cone_cylinder	#42 Additive Inverse	#43 Multiplicative Inverse	#44 Rational and Irrational numbers
1	D	C	D	B
2	D	D	A	B
3	D	C	C	A
4	C	B	B	D
5	A	A	D	A
6	B	C	C	C
7	B	D	A	C
8	B	A	A	D
9	C	B	D	B
10	B	B	B	A
11	C	C	A	A
12	D	D	C	C
13	A	A	D	B
14	D	B	B	D
15	A	A	C	D

Pre-Algebra Vol 2

KEY

Q #	#45 Exponents	#46 Slope (2 points)	#47 Slope Intercept	#48 Slope graph
1	A	D	D	D
2	D	B	A	A
3	B	B	D	C
4	A	D	A	A
5	C	D	D	C
6	A	D	C	D
7	D	D	B	A
8	B	A	C	D
9	D	C	A	D
10	B	A	B	C
11	C	C	B	A
12	A	C	A	C
13	B	C	C	B
14	B	A	C	C
15	D	B	C	B

KEY

Q #	#49 Slope of the straight line	#50 Parallel line slope	#51 Perpendicular line slope	#52 Missing angle
1	A	A	A	D
2	D	D	A	D
3	A	C	A	B
4	A	C	D	A
5	B	D	B	D
6	A	C	D	B
7	C	A	B	A
8	A	B	A	D
9	C	C	B	A
10	C	D	C	B
11	C	B	A	A
12	C	D	C	B
13	D	B	C	C
14	C	B	A	D
15	B	B	C	A

Pre-Algebra Vol 2

KEY

Q #	#53 Missing angle	#54 Reflections	#55 Rotations	#56 Translations
1	A	C	C	D
2	C	A	A	A
3	A	C	C	B
4	D	D	B	A
5	A	D	A	D
6	B	A	B	B
7	A	D	A	D
8	A	A	D	B
9	D	D	C	A
10	C	D	B	B
11	D	C	D	D
12	D	B	C	D
13	A	A	A	D
14	C	A	D	B
15	D	D	D	D

KEY

Q #	#57 One step word	#58 Simultaneous equations	#59 Simultaneous equations
1	C	A	B
2	D	A	B
3	C	C	D
4	B	D	B
5	D	D	A
6	A	B	D
7	A	B	D
8	D	A	C
9	C	C	A
10	C	D	B
11	B	A	D
12	B	B	B
13	A	D	D
14	D	C	D
15	A	B	A

#60 Line graph - Answer keys

1. C	28. D
2. A	29. A
3. D	30. D
4. A	31. C
5. C	32. C
6. A	33. B
7. D	34. C
8. B	35. A
9. D	
10. B	
11. C	
12. D	
13. A	
14. B	
15. A	
16. C	
17. B	
18. A	
19. C	
20. A	
21. B	
22. A	
23. D	
24. D	
25. B	
26. C	
27. C	

#61 Scatter plots - Answer keys

1. C
2. B
3. B
4. B
5. C
6. A
7. C
8. A
9. A
10. A

#62 Stem leaf plot - Answer keys

1. C
2. A
3. D
4. A
5. A
6. D
7. C
8. A
9. C
10. A
11. A
12. D
13. A
14. D
15. B
16. A
17. C
18. A
19. D
20. A

#63 Dot plot - Answer keys

1. C
2. A
3. D
4. A
5. C
6. A
7. D
8. B
9. D
10. A
11. A
12. D
13. A
14. A
15. D
16. C

#64 Pie chart - Answer keys

1. A
2. B
3. A
4. D
5. C
6. C
7. B
8. A
9. C
10. D

Pre-Algebra Vol 2

Answer Keys

11. A
12. D
13. C
14. A
15. C
16. D
17. A
18. C
19. B
20. A
21. B
22. D
23. C
24. A
25. C
26. C
27. D
28. B
29. D
30. B
31. C
32. A
33. A
34. C
35. D

www.ingramcontent.com/pod-product-compliance
Lightning Source LLC
Chambersburg PA
CBHW082028120526
44592CB00038B/2234